T0293762

Genetic Diversity

Genetic Diversity

Editor: Hudson Moran

www.callistoreference.com

Callisto Reference,
118-35 Queens Blvd., Suite 400,
Forest Hills, NY 11375, USA

Visit us on the World Wide Web at:
www.callistoreference.com

ISBN: 978-1-64116-738-3 (Hardback)

Cataloging-in-publication Data

Genetic diversity / edited by Hudson Moran.
   p. cm.
Includes bibliographical references and index.
ISBN 978-1-64116-738-3
1. Variation (Biology). 2. Genetics. 3. Biodiversity.
4. Molecular evolution. I. Moran, Hudson.
QH401 .G46 2023
576.54--dc23

Table of Contents

Preface

In my initial years as a student, I used to run to the library at every possible instance to grab a book and learn something new. Books were my primary source of knowledge and I would not have come such a long way without all that I learnt from them. Thus, when I was approached to edit this book; I became understandably nostalgic. It was an absolute honor to be considered worthy of guiding the current generation as well as those to come. I put all my knowledge and hard work into making this book most beneficial for its readers.

Genetic diversity refers to the range of different inherited traits or the variation in the alleles and genotypes within a species. The genetic diversity of a particular habitat depends on factors such as the nature of the environment, which in turn consists of factors such as climate and the availability of food and other natural resources. Genetic diversity occurs due to three distinct causes, namely, mutation, recombination, and immigration of genes. It plays a crucial role in deciphering evolutionary and adaptive process for most species. It is also used to implement conservation strategies and crop management. Genetic diversity is gradually declining due to human activities. Around the world, forest habitats and wetlands are being altered, most of which are rich in flora and fauna species. Such lands are now being drained for agriculture and urban development. Humans are engaged in overfishing activities, which has led to the endangerment of many marine fish species and disruption of the ocean ecosystems. This book provides comprehensive insights on genetic diversity. A number of latest researches have been included to keep the readers up-to-date with the recent developments in the concerned area of study.

I wish to thank my publisher for supporting me at every step. I would also like to thank all the authors who have contributed their researches in this book. I hope this book will be a valuable contribution to the progress of the field.

Editor

1

Genetic Diversity of Mexican Avocado in Nuevo Leon, Mexico

Adriana Gutiérrez-Díez,
Enrique Ignacio Sánchez-González,
Jorge Ariel Torres-Castillo,
Ivón M. Cerda-Hurtado and
Ma. Del Carmen Ojeda-Zacarías

1. Introduction

Mexico is considered the center of origin of avocado, and in the mountains of Sierra Madre Oriental in Nuevo Leon the remains of an avocado have been found, which are evidence of the place or origin of *Persea americana* var. *drymifolia*, or the Mexican avocado. Wild creoles have been found growing with native vegetation and have contrasting characteristics with cultivated varieties (improved creoles) and creoles (seedling) found in orchards. All this diversity represents a valuable source of genes and genetic combinations that can be used in *Persea* breeding programs. Therefore, we performed studies in the north and south areas of Nuevo Leon, Mexico, in order to determine the genetic diversity of this crop and implement actions for the characterization and conservation of this genetic resource, which has been disappearing due to changes in soil use, the introduction of improved varieties and the destruction of their habitat.

An analysis of 42 Mexican avocado samples based on the fruits traits as: fruit weight, fruit length, fruit diameter, seed weight, seed length, seed diameter, length seed cavity, the ratio of fruit length/fruit diameter and the ratio of seed weight/fruit weight, was carried out. The results of the analysis showed that the classification of genetic diversity with these fruit traits was not well represented. The characterization of the samples with AFLP molecular markers, allowed the differentiation of five genotypes in the cluster analysis and the separation of those with the same local name, thus demonstrating genetic differences within Mexican avocado genotypes. When the molecular data were analysed with the morphological data, 11 genotypes were differentiated and the separation of genotypes with the same local name was possible.

At present, we are working with 10 genotypes to determine genomic regions that help us to identify and differentiate genotypes. The research results have identified about 37 improved creoles, 19 creoles (seedling) and nine wild creoles (seedling), all of them with different traits. To preserve this diversity, we are implementing a local germplasm bank together with Mexican avocado growers in order to start a breeding program.

2. Origin of Mexican avocado (*Persea americana* var. *drymifolia*)

Mexico has a wide variety of types of avocado-there are at least 20 different species related to the avocado [1]. The avocado is part of the Lauraceae family, considered by many botanists to be among the most primitive of the dicotyledonous plant; the centre of origin of the avocado is located in the highlands of central and east-central Mexico and the Guatemalan highlands [2]. In Mexico, three races are recognized: Mexican, Guatemalan and West Indian, which were classified as botanical varieties [3]; the possible centre of origin of these races is shown in Figure 1 [4].

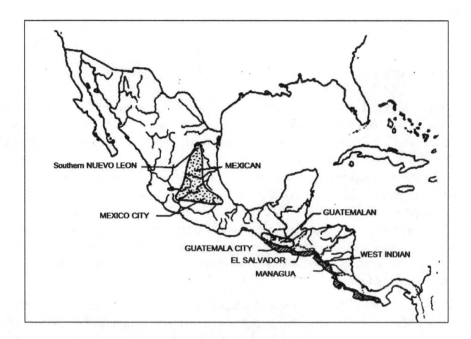

Figure 1. Possible origin centres of the three avocado races [4].

Persea americana var. *drymifolia*, or the Mexican avocado, is the oldest variety used as food [2]; seeds of avocados were found in cave deposits (El Riego, Purrón and Coxcotlán) in the Tehuacan Valley, Puebla in Mexico, the oldest cotyledon from the Coxcotlán cave deposit is dated to at least 7,000 B.C. [4]. In 1966, the geographical distribution of the *P. americana* var. *drymifolia* in Mexico was determined in the mountain forests of the eastern and south-eastern and in the Pacific zone (Figure 2) [5]. The findings of Mexican primitive avocados in the Sierra Madre Oriental of Nuevo Leon show that this area is part of the centre of origin of *Persea americana* var. *drymifolia* [6].

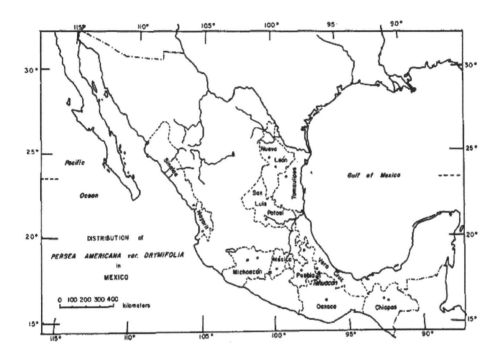

Figure 2. Geographical distribution of *Persea americana* var. *drymifolia* in Mexico [5].

The Mexican avocado is grown in different intensive agricultural systems and both in orchards and backyard gardens, these forms of production are important centres of experimentation, plant introduction, empirical improvement and shelters of unique genetic diversity that contain genes that have not been studied and could have potential use in breeding pro-grammes. *Persea americana* var. *drymifolia* trees are characterized by their resistance to cold and high oil content, a very distinctive characteristic is its strongly aromatic leaves (anise scented) in almost all individuals; usually, they grow at altitudes greater than 2,000 m. The leaves are dark green with light green or reddish young shoots; the fruit has a thin, smooth and soft skin, the seed can be adhered or loose, the cotyledons are smooth or slightly rough, and it is common to find fibre in the flesh, although this is not found in most cultivated species [1].

Given that the avocado is an open-pollinated species, it contains great genetic variability with almost unlimited possibilities for utilization [7]; a wide diversity of germplasm allows the advancement of botanical and agronomic knowledge and the development of new cultivars [6]. The germplasm of Mexico has been the basis of breeding programmes in other countries, so it is necessary to focus on the exploration, classification and preservation of the germplasm of the genus *Persea* to implement breeding programs. The generation of improvement varieties involves searching for genes that may exist in some cultivated varieties or wild plants; if these sources of genes do not exist, the possibility of the transferring of its characteristics will be lost forever.

In Nuevo Leon, Mexico, it is still possible to find different wild plants of the Mexican avocado (wild creoles) growing among natural vegetation, and their morphological traits are contrasting with cultivated varieties. Cultivated varieties consist of local selections of plants that have been grown for several years and which the growers have selected based on their production and quality (fruit size, mainly); these cultivars are grafted trees with genotypes of interest and are called 'improved creoles'. Native plants are those from avocado seeds that have not been grafted (seedlings) and are called 'creoles'. The genetic diversity of wild plants, improved creoles and creoles, represents a valuable source of genes and genetic combinations that can be used in *Persea* breeding programmes.

The main problems in identifying the genetic variability of Mexican avocados in Nuevo Leon, Mexico, are, among other factors, the diversity of local names that are used for the improved creoles, and the use of the same rootstock for grafting different varieties, resulting in a tree with branches belonging to different cultivars (Figure 3). Since 2005, in order to study this genetic diversity, we performed studies of the morphological and molecular characterization of Mexican avocado varieties.

3. First review of *Persea americana* var. *drymifolia* in Nuevo Leon, Mexico

In order to identify the genotypes of Mexican avocados (improved creoles, creoles and wild creoles), from June 2005 to November 2006, surveys were conducted in the municipalities of Dr. Gonzalez and Sabinas Hidalgo, located north of Nuevo Leon, Mexico, and in the municipalities of Aramberri, General Zaragoza and Rayones, located south of the state. A total of 30 improved creoles and 19 creoles and wild creoles, were identified. We collected five fruit of each of the trees that had fruit at the time of the visit (26 improve creoles and 12 creoles). The improved creoles collected in Dr. Gonzalez were Huevo de Toro Blanco, Larralde, Verde Perez and Perales, whereby the fruits of these cultivars mature in green color; and Anita, Floreño, Huevo de Toro, El Cuervo, Negro Santos Normal, Negro Santos Especial, Rodriguez and Rosita, the fruits of these improved creoles mature in black color (Figure 4), also were collected four creoles (Figure 5). In Aramberri, we collected the improved creoles: Mantequilla, Pagua, Platano Delgado, Maria Elena, Fuerte, Leonor, Pato and De la Peluqueria (Figure 6), as well as five creoles (Figure 5)-in the case of Mantequilla, the skin of the ripe fruit is yellow. In the General Zaragoza municipality, only two creoles were identified (Figure 5 and Figure 8). In Sabinas Hidalgo, improved creoles were identified, such as Anita, Floreño, Negro Santos and Pepe (Figure 7). In Rayones, the improved creoles were Verde Fuerte and Pagua (Figure 7), as well as two creoles (Figure 5). Four improved creoles had no fruits: Lampazos, Blanco and Sanjuanero of Dr. Gonzalez municipality, and Israel of Aramberri municipality.

Figure 3. Mexican avocado tree, A) rootstock, B) Platano Grueso improved creole.

Fourteen improved creoles were reported in Sabinas Hidalgo. In the study of evaluation of native avocados in the northern region of Nuevo Leon [8], six genotypes corresponded with those we found. In this same report, 10 improved creoles that were not documented by us were reported in Sabinas Hidalgo: Blanquito, Fosa, Cuervo, Pera, Sabroso, Chapeño, Pecoso, Pila, Salazareño, Especial. Negro Santos and Especial genotypes were reported in Bustamante [8]- this area was not considered in our study.

El Salto is a wild creole collected in the municipality of General Zaragoza and it was named this way because the trees were located in Parque Recreativo El Salto, located in this municipality. The fruit has a length of 2 cm and, according to researchers at Red Aguacate-SNICS-SAGARPA in Mexico, it could be a primitive Mexican avocado genotype-this fruit's characteristics are similar to those reported by Storey, who mentioned the discovery of a primitive form of avocado fruit which is about 2 cm length, in the pine and oak forests of Nuevo Leon, Mexico [4]; this vegetation type coincides with the area where we located this genotype (Figure 8).

The harvest of Mexican avocados in Nuevo Leon, Mexico, is performed during the period from June to December, although there are genotypes with fruit production throughout the year.

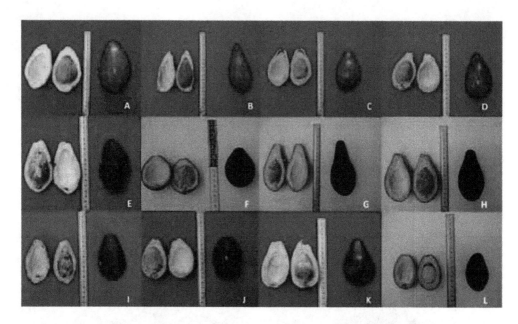

Figure 4. Mexican avocado improved creoles of Dr. Gonzalez, Nuevo Leon, Mexico. A) Huevo de Toro Blanco, B) Larralde, C) Verde Perez, D) Perales, E) Anita, F) Floreño, G) Huevo de Toro, H) El Cuervo, I) Negro Santos, J) Negro Santos Especial, K) Rodriguez, L) Rosita.

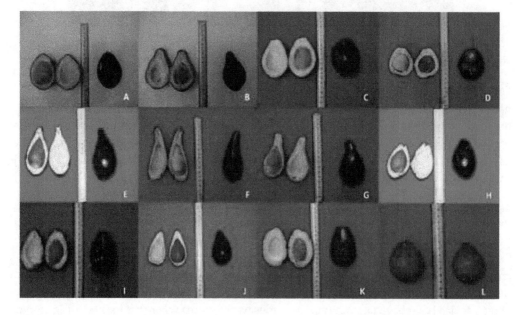

Figure 5. Creoles (seedlings) of *P. americana* var. *drymifolia*. A), B), C) and D) Dr. González; E), F), G), H) and I) Aramberri; J) General Zaragoza; K) Sabinas Hidalgo; L) Rayones.

Figure 6. Mexican avocado improved creoles of Aramberri, Nuevo Leon, Mexico. A) Pagua, B) Maria Elena, C) Leonor, D) Pato, E) Mantequilla, F) Platano Delgado, G) De la Peluqueria.

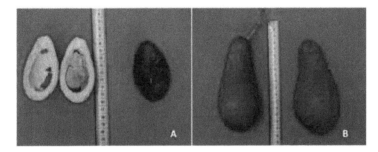

Figure 7. Mexican avocado improved creoles of Sabinas Hidalgo and Rayones, Nuevo Leon, Mexico. A) Pepe (Sabinas Hidalgo); B) Verde Fuerte (Rayones).

Figure 8. Wild creole El Salto. A), B) fruits; C), D) trees.

4. Molecular and morphological characterization

In order to estimate the genetic diversity through morphological and molecular characteristics, in 2007 we worked with 42 trees of *P. americana* var. *drymifolia* of Aramberri and General Zaragoza, in Nuevo Leon, Mexico (Table 1).

Creole improved/creole	ID
Huevo de Paloma	P1
Campeon	P2, P6
Pagua	P3, P24
Platano Grueso	P4, P26
Platano Delgado	P7, P11, P12, P13, P14, P15, P21
Verde	P9
Maria Elena	P19, P33
Platano	P20, P27, P28, P29, P30, P38
Calabo	P23
Leonor	P25
Chino	P32
Creole	P5, P16, P18, P39, P44, P59, P60
Wild creole	P45, P46, P47, P48, P49, P50, P53, P56, P57

Table 1. Identification of Mexican avocados samples used to determine morphological and molecular diversity.

The collections of the samples were conducted both in gardens and in areas of wild vegetation. The identification of the samples was performed according to the local name provided by the growers-those samples of non-grafted trees were identified as creoles and samples collected in areas of wild vegetation were identified as wild creoles. Eleven improved creoles corresponding to 26 samples, seven creoles and nine wild creoles were identified (Table 1). The Platano Grueso improved creole is more commercially accepted by the characteristics of size and appearance of the fruit; this cultivar characteristically presents an early harvest starting cycle, in mid-June, which limits its production from concentrating in a relatively short period of time [9]. The tendency of some producers is the replacement of treetops to grafts of this cultivar. Some other cultivars are preserved due to its adaptive characteristics, palatability and harvest out of season, allowing marketing for the best price. The study of the evaluation of creole avocados in the southern region of the state of Nuevo Leon reported 23 improved varieties [9], of which 13 coincide with those that have been identified in our studies.

Due to the lack of definition in the branching pattern of trees by grafting different varieties in the same plant (Figure 3), it was decided only to perform a morphological evaluation of the fruits. For the molecular analysis of the samples and the morphological analysis of their fruits, 10 leaves and five ripe fruits per tree were collected. The characterization of the fruits was performed according to the morphological descriptors for avocado fruit given by the International Plant Genetic Resources Institute [10]. The characteristics were: fruit weight (g), fruit length (cm), fruit diameter (cm), seed weight (g), seed length (cm), seed diameter (cm), seed cavity length (cm), seed cavity diameter (cm), ratio of the fruit length/fruit diameter, and ratio of the seed weight/fruit weight. In order to determine the variables with greater weight on the morphological characterization and the behaviour of each variable, the mean, standard deviation and coefficient of variation were estimated. The data analysis was performed by principal components analysis and based on this; a cluster analysis was performed using the UPGMA method (un-weighted pair groups method with arithmetic averages) with the Gower distance (1-S) [11]. Statistical analyses were performed using the software InfoStat/Pv.2006p.3 [12].

In the fruit's morphological characterization, the traits with higher coefficients of variation were: seed weight (39.42%) and fruit weight (38.42%); the seed cavity length and fruit length showed coefficients of 24.82% and 20.34%, respectively. Values greater than 50% suggest that the characteristics have the highest variability within a species, while values below 20% indicate little variability in the species [11], so these characteristics were considered as classificatory.

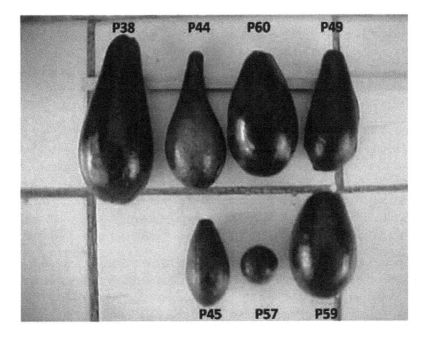

Figure 9. Comparison of the sizes and shapes of fruits of *Persea americana* Mill. var. *drymifolia* from Aramberri and General Zaragoza, Nuevo Leon, Mexico [13]. The sample identification is according to Table 1.

As to the seed weight and fruit weight, the highest averages were 69.76 g and 251.40 g, respectively, corresponding to P26 (Platano Grueso); the lowest values for this variable were

7.08 g and 10.53 g, respectively, corresponding to P57 (El Salto). In the case of seed cavity length, the highest average corresponded to P21 (Platano Delgado) with 11.46 cm, the lowest average was 2.60 cm and corresponded to P57 (El Salto). For fruit length, the highest average value was 12.76 cm and corresponded to P4 (Platano Grueso), the lowest value corresponded to El Salto and was 2.88 cm; the fruit traits of this wild creole contrast sharply with the others samples collected (Figure 9).

The relationship between the traits evaluated and the similarity between the samples was performed by principal component (PC) analysis with standardized data. The PC1 and PC2 represented 75.8% of the total variance, values greater than 60-70% explaining a reasonable percentage of the total variability of the samples [14]. The traits with the greatest influence in PC1 (59.7%) were fruit weight, seed weight and fruit length, while in PC2 (16.1%) they were seed diameter and seed cavity length. The cophenetic correlation was 0.984; PC analysis results coincided with the results of analysis of variation coefficients; of the samples organized in six clusters, P57 had the highest distance, and P45 and P48 (wild creoles) formed a cluster. P53, P56 and P11 formed another cluster-the first two corresponding to wild creoles while P11 corresponds to Platano Delgado. The samples P25 (Leonor) and P21 (Platano Delgado) were separated from the rest of the samples, while the other samples were grouped into a cluster.

The dendrogram was performed with the following variables: seed weight, fruit weight, seed cavity length and fruit length; with the addition of the ratio of fruit length/fruit diameter and the ratio of seed weight/fruit weight, the cophenetic correlation increased to 0.888. Four clusters were formed at a distance of 0.23; two clusters coincide with those formed by the principal components analysis. Cluster I was formed with P57, cluster II with P21. Cluster IV was formed by five samples identified as wild creoles (P45, P47, P48, P53 and P56), sample P16 (creole), P1 (Huevo de Paloma), P25 (Leonor) and P32 (Chino). Cluster III was formed with the remaining samples (Figure 10).

Although the morphological differences between fruits are evident, only the sample P57 was identified. Linkage distances within the dendrogram reflected the differences and inconsistencies between presumably identical genotypes; we expected that the samples identified with the same name would be grouped together, which did not happen. These results agree with those reported by [15], who assert that fruit size is a trait that does not help in the differentiation of wild and cultivated avocado plants, because trees produce fruits of different sizes. Avocado groups are not well represented when the morphological characteristics of fruits are used [16].

Traditional plant identification is performed by phenotypic characterization-this is slow and limited, and the expression of the quantitative characteristics is subject to environmental influences. Molecular markers allow the identification, classification and use of genetic diversity present in the genomes of plants; differences or similarities at the DNA level in the individuals are observed directly. The AFLP (amplified fragment length polymorphism) is a highly efficient molecular marker [17]-in avocados they have been used to study the genetic diversity of germplasm [16, 18, 19].

The molecular characterization of the samples was performed using AFLP molecular markers; AFLP generation was performed using the IRDye Fluorescent AFLP Kit for Large Plant

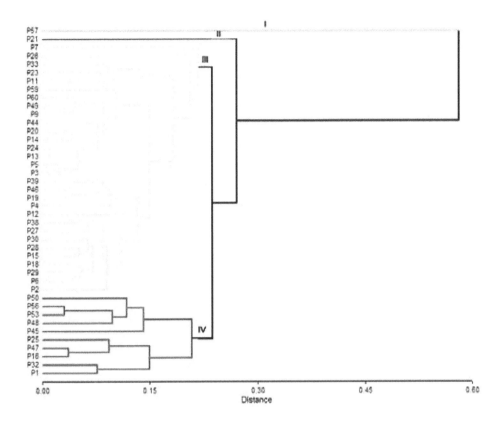

Figure 10. Dendrogram of the fruit's morphological characteristics of 42 *Persea americana* Mill. var. *drymifolia*.

Genome Analysis (LI-COR® Biosciences); for the construction of the binary data matrix, it was assumed that bands with the same molecular weight are identical, assigning the number 1 to the presence of bands and the number 0 to the absence of bands [13]. Similarity indices were determined by the Gower coefficient (1-S). A cluster analysis of the molecular data was performed using the UPGMA method using standardized variables through the Info/Gen 2006p.1v software [20]; the cophenetic correlation coefficient was calculated.

With the cluster analysis of 683 AFLP markers (Figure 11), two clusters were defined (distance 0.14). The samples P9 (Verde), P44 and P39 (Creoles), P38 and P29 (Platano), P4 (Platano Grueso), P3 (Pagua) and P23 (Calabo), were not clustered. The separation of these samples shows that they are different genotypes, but in those samples that were not clustered with samples with the same local name, the question is whether there is genetic variation among the improved creoles. Cluster "I" was composed of samples P48, P46 and P45, identified as wild creoles, while cluster "II" was composed of 32 samples. By decreasing the distance to 0.09 in the dendrogram, samples identified as Platano (P30, P15, P13, P12) and Platano Delgado (P11), were grouped; while P56, P50 (both wild creoles) and P32 (Chino) samples were separated. The clustering at very short distances (0.015) is between P6 (Campeon), P7 (Platano Delgado) and P1 (Huevo de Paloma), and between P53 (wild creole) and P33 (Maria Elena), which shows that they are closely related. The clustering at distances greater than samples identified with the same name calls into question the genetic identity of improved creoles.

To perform the comparative analysis between cluster groups formed in the morphological and molecular analysis, a mixed data matrix was constructed with morphological data and AFLP binary data; a cluster analysis was performed by UPGMA with the Gower distance (1 S) through the software Info/Gen 2006p.1v [20]. As in previous cases, the cophenetic correlation coefficient was calculated.

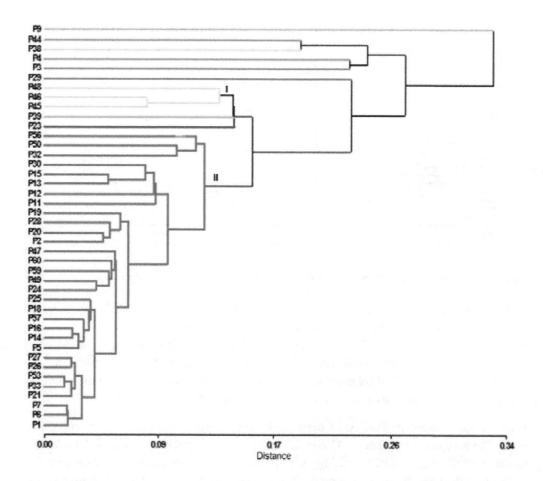

Figure 11. Dendrogram of molecular data (AFLP markers) of 42 *Persea americana* Mill. var. *drymifolia*.

The cluster analysis of the molecular and morphological data showed one cluster (I) formed by 31 samples (distance of 0.13), while 11 samples showed no clustering pattern: P48 (wild creole), P57 (wild creole), P45 (wild creole), P9 (Verde), P44 (creole), P4 (Platano Grueso), P3 (Pagua), P56 (creole), P50 (wild creole), P38 (creole) and P29 (Platano) (Figure 12). The separation of P9, P44, P3 and P4 coincided with that obtained in the molecular data analysis. The separation of P48, P57, P45, P56 and P50 identified as wild creole is a difference between the analysis of mixed data and analysis of molecular data. With increasing distance in the dendrogram, the P57 and P45 (wild creoles) samples were clustered; these samples are characterized by the lowest values of fruit length and seed length. When the cluster distance was increased in the dendrogram, the influence of the morphological variables in the formation of the groups was evident, so the group consisting of P48, P57 and P45 was characterized by the lowest seed weight values. When analysing in detail the scheme of clustering, we observed that P26 and P27 were grouped at a shorter distance-an important feature of these samples is that they showed the highest fruit weight and that they correspond to the improved creoles Platano Grueso and Platano, respectively. As such, we can conclude that they present similarities.

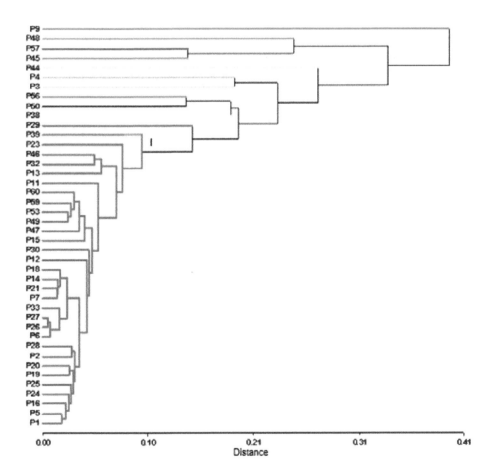

Figure 12. Dendrogram of morphological characteristics and molecular data (AFLP markers) of 42 *Persea americana* Mill. var. *drymifolia*.

The results of this study demonstrated that AFLP molecular markers allow the estimation of the genetic diversity of the Mexican race of *Persea americana* Mill. when molecular data are combined with morphological data. When only molecular data generated by AFLP are used, the differentiation between genotypes was ambiguous; we demonstrated the effectiveness and facility of AFLP used to characterize avocado accessions based on the race origin [19]; however, there is no reference about its usefulness in differentiating between varieties.

5. *In vitro* culture callus

We have worked with *in vitro* culture techniques to conserve the germplasm of different avocado genotypes. The selection of the explant to be used in the conservation of plant tissue depends on the resources and the goal of the project; in avocados, different explants have been used to obtain different types of morphogenesis [21]. The avocado morphogenetic capacity under *in vitro* conditions is linked to the use of materials with juvenile characteristics [22]. The avocado cotyledon callus was kept for over 15 years without apparent differentiation conditions [23]-these tissues remained as callus divisions through a series of small segments and were transferred to fresh media at intervals of one-to-three months. The callus tissue is essential for obtaining somatic embryos by indirect embryogenesis [24, 25]. Avocado somatic embryos were regenerated from calluses [26-29]. We used the developed young leaves of a Mexican avocado tree (six years old) as explants to establish an *in vitro* culture callus.

The avocado leaves were washed with a commercial detergent and running water for 30 minutes. Subsequently, petiole were cut and the leaves were placed in an antioxidant solution, 400 mg L^{-1} of ascorbic acid, 150 mg L^{-1} of citric acid and 30 g L^{-1} of sucrose with 1 g L^{-1} systemic fungicide metalaxyl-M (Ridomil, Syngenta Agro), for 2 min at a vacuum pressure of-20 bar; they were placed in alcohol 70% v / v for 30 s followed by three rinses with sterile water, then placed in a solution of 2.4% NaOCl with two drops of Tween 20 for each 100 ml for 15 min, followed by three rinses with sterile distilled water; the leaves were kept in sterile antioxidant solution above without systemic fungicide, until their establishment *in vitro*. Explants of 1 cm^2 were obtained and placed in glass vials with 20 ml of DCR medium [30] pH 7.5, supplemented with vitamins and growth regulators, 200 mg L^{-1} of inositol, 50 mg L^{-1} of glutamine, 500 mg L^{-1} of casein hydrolyzate, 0.3 mg L^{-1} of 6-benzylaminopurine, 0.01 mg L^{-1} of naphthaleneacetic acid, 0.1 mg L^{-1} of thiamine, 0.5 mg L^{-1} of nicotinic acid, 0.5 mg L^{-1} of pyridoxine, 2 mg L^{-1} of glycine, 10 mg L^{-1} of ascorbic acid, 10 mg L^{-1} of citric acid, 1.5 mg L^{-1} of 2,4-D, 0.3 mg L^{-1} of kinetin, 20 g L^{-1} of sucrose and 4.5 g L^{-1} of Phytagel. The glass vials with the explants were placed in a controlled temperature of 20 ° C ± 2 ° C in complete darkness for callus formation. Callus formation from the leaves was completed out in 30 days (Figure 13).

Using the same culture medium, two subcultures of each callus were performed with a period of 30 to 45 days between them. AFLP markers were generated and 341 polymorphic bands were used to calculate the polymorphic information content (PIC) and the genetic distance to measure genetic stability of the avocado callus. We analysed molecularly 94 samples, the mother plant, 31 calluses formed from the mother plant explants, 31 calluses from the first subculture, and 31 calluses from the second subculture-the results of this research are in press.

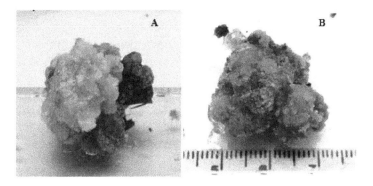

Figure 13. A) Avocado callus from leaves, B) avocado callus from the second subculture.

6. Present investigations

Actually, we are working with the fruit gene expression of genotypes with contrasting characteristics (Figure 14). Using the technique of differential display [31], we obtained differential fragments of the genome of the fruit (Figure 15); these amplified fragments were sequenced in order to identify regions of the genome to help us in the identification and differentiation of genotypes (improved creoles, creoles and wild creoles). We are analysing the sequences obtained with data from the NBCI and hope that these results can be used in the future in the genetic characterization of genotypes as well as in avocado breeding programs.

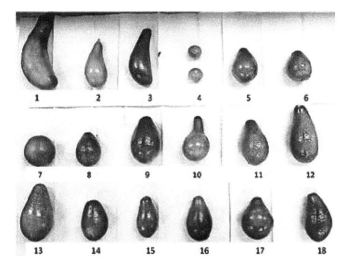

Figure 14. Fruits of *Persea americana* Mill. var. *drymifolia* of Aramberri and General Zaragoza, Nuevo Leon, Mexico. 1) Pato, 2) Amarillo, 3) Cuerno, 4) El Salto (wild creole), 5) Huevo de Paloma, 6) Mantequilla, 7) Criollo Bola (creole), 8) Pagua, 9) Platano, 10) and 14) Criollos (creoles), 11) Campeon, 12) Platano Delgado, 13) Platano Grueso, 15) Criollo Todo el Año (creole), 16) Maria Elena, 17) Leonor, 18) Salvador.

In order to preserve all the genetic variability of the Mexican avocado that has been detected in Nuevo Leon, Mexico, a local germplasm bank is being implemented together with growers where they can safeguard copies of improved creoles, creoles and wild creoles. The intention is to have plant material for them and to start in a short period with the creation of a breeding program for the Mexican avocado.

7. Conclusion

The genetic diversity of *Persea americana* var. *drymifolia*, found in Nuevo Leon, Mexico, is broad and can be used for incorporation into breeding programme cultivars, as well as for use as rootstocks and inter-stocks; wild varieties are a genetic patrimony that can provide innovative sources of advantageous use for producers. Identifying the characteristics of agronomic interest (genes) boosts the Mexican avocado crop through its alternate use (processed fruit, oil extraction, medicinal use, use as a condiment, animal feed, etc.). The classification and conservation of plant germplasm should be seen as an activity to preserve this heritage of diversity-it requires a major effort to preserve the genotypes of cultivated plants, but especially wild and native plants that are threatened with the destruction of their natural habitat and being replaced by improved varieties.

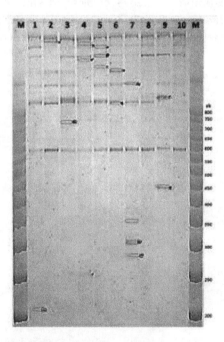

Figure 15. Differential fragments of *Persea americana* var. *drymifolia* fruits. Lanes: M, molecular weight marker, 1) Pato, 2) Cuerno, 3) El Salto, 4) Mantequilla, 5) Criollo Bola, 6) Criollo 2, 7) Platano Grueso, 8) Criollo Todo el Año, 9) María Elena, 10) Leonor. Polyacrylamide gel 6%.

Author details

Adriana Gutiérrez-Díez[1*], Enrique Ignacio Sánchez-González[1], Jorge Ariel Torres-Castillo[2], Ivón M. Cerda-Hurtado[1] and Ma. Del Carmen Ojeda-Zacarías[1]

*Address all correspondence to: mcgudiez@aol.com

1 Universidad Autonoma de Nuevo Leon. Faculty of Agronomy. General Escobedo, Nuevo Leon, Mexico

2 Universidad Autonoma de Tamaulipas. Institute for Applied Ecology. Ciudad Victoria, Tamaulipas, Mexico

References

[1] Barrientos-Priego A.F., López-López L. Historia y Genética del Aguacate. In: Memoria de la Fundación Salvador Sánchez Colín. Centro de Investigaciones Científicas y Tecnológicas del Aguacate en el Estado de México; 2002. pp. 100-121.

[2] Williams LO. The Botany of Avocado and its Relatives. In: Sauls JW, Phillips RL, Jackson LK. (eds.) Proceedings of the First International Tropical Fruit Short Course: The Avocado, 5-10 Nov 1976, Gainsville, Florida, United States; 1976.

[3] Bergh B., Ellstrand N. Taxonomy of the Avocado. In: California Avocado Society Yearbook 70. 1986. pp. 135-146.

[4] Storey WB., Bergh B., Zentmyer GA. The Origin, Indigenous Range, and Dissemination of the Avocado. In: California Avocado Society Yearbook 70. 1986. pp. 127-133.

[5] Smith CE. Archaeological Evidence for Selection in Avocado. Economic Botany 1966;20 169-175.

[6] Sánchez-Pérez JL. Recursos Genéticos de Aguacate (*Persea americana* Mill.) y Especies Afines en México. Revista Chapingo Serie Horticultura 1999;5 7-18.

[7] Bergh BO. The Origen, Nature, and Genetic Improvement of the Avocado. In: California Avocado Society Yearbook 76. 1992. pp. 61-75.

[8] Acosta E, Almeyda IH, Hernández I. Evaluation of Native Avocados in the Northern Region of Nuevo Leon, Mexico. Revista Mexicana de Ciencias Agrícolas 2013; 4(4) 531-542.

[9] Acosta E, Hernández I, Almeyda IH. Evaluation of Creole Avocados in Nuevo Leon, Mexico: Southern Region. Revista Mexicana de Ciencias Agrícolas 2012; 3(2) 245-257.

[10] IPGRI. Descriptors for Avocado (*Persea* spp.). International Plant Genetic Resources Institute; 1995.

[11] Franco TL., Hidalgo R. (eds.). Análisis Estadístico de Datos de Caracterización Morfológica de Recursos Fitogenéticos. International Plant Genetic Resources Institute; 2003.

[12] InfoStat. InfoStat versión 2006. Universidad Nacional de Córdova; 2006.

[13] Gutiérrez-Díez A, Martínez-De la Cerda J, García-Zambrano E, Iracheta-Donjuan L, Ocampo-Morales JD, Cerda-Hurtado IM. Study of Genetic Diversity of Native Avocado in Nuevo Leon, Mexico. Revista Fitotecnia Mexicana 2009;32(1) 9-18.

[14] Balzarini M, Arroyo A, Bruno C, Di Renzo J. Análisis de Datos de Marcadores con Info-Gen. In: XXXV Congreso Argentino de Genética, 2006, San Luis, Argentina; 2006.

[15] Gama-Campillo L, Gomez-Pompa A. An Ethnoecological Approach for the Study of *Persea:* A Case Study in the Maya Area. In: Lovatt C, Holthe PA, Arpaia ML (eds.) Proceedings of the Second World Avocado Congress, 21-26 April 1991, California, United States; 1992.

[16] Rodríguez N, Rhode W, González-Arencibia C, Ramírez-Pérez IM, Fuentes-Lorenzo JL, Román-Gutiérrez MA, Xiqués-Martín X, Becker D, Velásquez-Palenzuela JB. Caracterización morfológica, bioquímica y molecular de cultivares de aguacatero (*Persea americana* Mill.) en Cuba. In: Proceedings of V World Avocado Congress 2003, 19-24 October 2003, Granada-Málaga, Spain; 2003.

[17] Vos P, Hogers R, Bleeker M, Reijans M, van de Lee T, Hornes M, Fritjers A, Pot J, Peleman J, Kuiper M, Zabeau M. AFLP: A New Technique for DNA Fingerprinting. Nucleic Acids Research 1995;23 4407-4414.

[18] Gallo-Llobet L. Sustainable Agriculture: the Role of Integrated Management or Root Rot (*Phytophthora cinnamomi* Rands) in Avocado (*Persea americana* Mill.). In: Summary Reports of European Commission Supported STD-3 Projects (1992-1995), Technical Centre for Agricultural and Rural Cooperation; 1999.

[19] Chao CT, Barrientos-Priego AF, Reyes-Alemán JC, Devanand PS. Relaciones Genéticas entre Accesiones de Aguacate de California y de México, Caracterizadas por Marcadores AFLP. In: Proceedings of V World Avocado Congress 2003, 19-24 October 2003, Granada-Málaga, Spain; 2003.

[20] Balzarini M, Di Renzo J. Info-Gen: Software para Análisis Estadístico de Datos Genéticos, Universidad Nacional de Córdoba, Argentina; 2003.

[21] Yassee MY. Morphogenesis of Avocado *in vitro*. A review. In: California Avocado Society Yearbook 77. 1993. pp. 101-105.

[22] Pliego-Alfaro AF, Barcelo M, López C. Rejuvenecimiento de una Planta Leñosa Frutal: Aguacate, y su Propagación por Cultivo de Tejidos. Expociencia 1986;97-103.

[23] Vidales FI. Efecto de los Reguladores de Crecimiento en los Procesos de Organogénesis y Embriogenesis Somatica de Aguacate *(Persea americana* Mill.). D.C. Tesis). Universidad de Colima, Mexico; 2002.

[24] Schroeder CA. Responses of Avocado Stem Pieces in Tissue Culture. In California Avocado Society Yearbook 60. 1976. pp. 160-163.

[25] Skeney KG, Barlass M. *In vitro* Culture of Abscised Immature Avocado Embryos. Annals of Botany 1983;52(5) 667-672.

[26] Mooney PA, Van Staden J. Induction of Embryogenesis in Callus from Immature Embryos or *Persea americana.* Canadian Journal of Botan 1987;65 622-626.

[27] Pliego F, Murashige T. Somatic Embryogenesis in Avocado *(Persea americana* Mill.) *in vitro.* Plant Cell Tissue and Organ Culture 1988;12 61-66.

[28] Efendi D, Litz RE. Cryopreservation of Avocado. In: Proceedings of V World Avocado Congress 2003, 19-24 October 2003, Granada-Málaga, Spain; 2003.

[29] Vidales FI, Salgado R, Gómez MA, Ángel E, Guillén H. Embriogénesis Somática de Aguacate *(Persea americana* Mill. cv. Hass In: Proceedings of V World Avocado Congress 2003, 19-24 October 2003, Granada-Málaga, Spain; 2003.

[30] Gupta PK, Durzan DJ. Shoot Multiplication from Mature Trees of Douglas-fir *(Pseudotsuga menziessi)* and Sugar Pine *(Pinus lambertiana).* Plant Cell Report 1985; 4 177-179.

[31] Liang P, Pardee AB. Differential Display of Eukaryotic Messenger RNA by Means of the Polymerase Chain Reaction. Science 1992;257(5072) 967-971. DOI:10.1126/science. 1354393.

2

Molecular Approach of Seagrasses Response Related to Tolerance Acquisition to Abiotic Stress

A. Exadactylos

1. Introduction

The debate surrounding climate change and its adverse effects on marine ecology is one of the most highly charged issues throughout the scientific community (e.g. Costanza et al., 1997; O'Neill, 1988). As far as seagrasses monitoring process is concerned, scientific data is needed that would contribute to the enhancement of marine environmental protection and their species conservation. Their use as biomarkers (Ferrat et al., 2003) is deemed as crucial due to the fact that they could be a reliable tool for researchers in the assessment of marine ecological status (transitional and coastal waters) in compliance with the Water Framework Directive (WFD, 2000/60/EC) and Marine Strategy (2008/56/EC) issued by the European Commission. Additionally, a challenge would be to deal with questions which arise from the underpinning tolerance mechanisms of seagrasses and whether they possess a sufficiently adjustable genetic background which in parallel can evolve in accordance with global warming.

Seagrasses play a critical role in the maintenance of marine environmental quality, creating complex, mosaic type habitats with high ecological and economic significance (Wiens et al., 1993; Hughes et al., 2003; Torre-Castro and Rönnbäck, 2004). The value of their contribution to the ecosystem is estimated at approximately 12,000€ per hectare/year, a part of which, concerns the support of commercial fish supplies (nurseries) and in general the conservation of marine biodiversity. Moreover, they contribute to coastal protection from sea waves, to the withholding of sediments and the recycling of nutritious substances (nutrient retention) (Cabaço et al., 2010), while they constitute important sources of carbon dioxide uptake from the atmosphere.

Seagrasses are highly productive submersed marine angiosperms that grow in shallow coastal and estuarine waters, providing key habitants of important ecological and financial value (Heck et al., 2003; Bloomfield and Gillanders, 2005; Heck et al., 2008). However, substantial declines in such habitats have been reported worldwide, mostly attributed to light reduction from algal overgrowth, sediment loading and re-suspension, anthropogenic disturbance and global climate change (Duarte and Prairie, 2005; Duffy, 2006; Orth et al., 2006; Burkholder et al., 2007; Leoni et al., 2008). Changes in sea level, fluctuations in salinity and temperature can alter seagrass distribution, productivity, and community composition (Short and Neckles, 1999; Alberto et al., 2008).

Angiosperms are a unique group of plants comprising more than 50 species of monocotyledons which have returned to the sea, while retaining numerous physiological and morphological characteristics of terrestrial plants (Arnaud-Haond et al., 2007; Ito et al., 2011; Rubio et al., 2011). In doing so, they have evolved in a medium with a much higher salinity than that tolerated by their terrestrial counterparts. However, our knowledge on salinity tolerance mechanisms in these marine plants is limited compared with that concerning terrestrial plants and marine algae (e.g. Vermaat et al., 2000; Torquemada et al., 2005; Hartog and Kuo, 2006; Waycott et al., 2006; Touchette, 2007).

Evolutionary studies of seagrasses, which reconstruct the origin and development of salinity tolerance in a variety of plant lineages, may help us to understand why artificial breeding has failed to produce robust and productive salinity tolerant crops. Such studies may also help us develop new salinity-tolerant lines by revealing the order of components acquisition on salinity tolerance, or indicating favorable genetic background on which salinity tolerance may be developed. By examining the repeated evolution of this complex trait we may identify particular traits, or conditions that predispose species to evolve a complex, multifaceted trait such as salinity tolerance and give rise to halophyte lineages. More generally, this may shed light on the adaptation of angiosperm lineages to extreme environments (Dupont et al., 2007; Sharon et al., 2009). In order to achieve these hypotheses, more information is required on, at a minimum, the effects of salinity on the growth and ion relations of a wider range of plant species that may prove to be seagrasses (Flowers, 2004).

Therefore, important questions could be posed: (i) whether all seagrass species tolerate salinity in, fundamentally, the same way; (ii) whether specific mechanisms can be identified and, if so, whether these are linked taxonomically; and (iii) whether specific mechanisms have evolved to deal with interactions between salinity and other environmental variables (Vicente et al., 2004; Flowers and Colmer, 2008; Wissler et al., 2011). If so, are these common to different taxonomic groups and how often has salinity tolerance evolved?

2. Review of literature

Seagrasses are monocot plants which have evolved from terrestrial ancestors that returned to the sea approximately 100 million years ago and have adapted to growing on the sea bed (Touchette and Burkholder, 2000). They are exposed to an inexhaustible source of K^+ and conditions that vary slightly from 11 mM K^+, 470 mM Na^+, and pH 8.2. Although cells of seagrasses have a normal physiology and are energized, as other plants, by an H^+-pump

ATPase (Fukuhara et al., 1996; Garciadeblas et al., 2001), their K^+ transport system must be adapted to living permanently in a medium with a high K^+ content.

In living cells, potassium (K^+) is the most abundant cation whose contribution is considerable due to its ability to maintain the electrical and osmotic equilibrium of cell membrane. Since K^+ was selected for these functions very early in the evolution of life, the cellular processes evolved within a K^+ rich medium and many of them became K^+-dependent. Plant cells are not exceptions to these K^+ requirements, but with the peculiarity that, in the Cambric Era, plants evolved on the rocks emerging from the sea where they had to adapt to taking up K^+ from an extremely poor environment. In these conditions, plants developed complex mechanisms of K^+ uptake and distribution. At present, most soils are less K^+ deficient than cambric rocks, but still K^+ occurs at low concentrations and K^+ acquisition and distribution play key roles in the physiology of contemporary terrestrial plants (Rodríguez-Navarro and Rubio, 2006).

In terrestrial plants, trans-membrane K^+ movements are mediated by several types of non selective cation channels (NSC) (Fig. 1), and by transporters that belong to two families KcsA-TRK and Kup-HAK, present in prokaryotes and eukaryotes (Schachtman and Schroeder, 1994; Quintero and Blatt, 1997; Santa-María et al., 1997; Fu and Luan, 1998; Kim et al., 1998; Rubio et al., 2000; Rodríguez-Navarro and Rubio, 2006). The extensive expression of KT-HAK-KUP transporters in many organs of the plant suggests that they coexist with K^+ channels and that their functions may be redundant or perhaps complementary to these channels (Garciadeblás et al., 2002). Low-affinity K^+ uptake is thought to be mediated primarily by K^+ channels whereas, high-affinity K^+ uptake is dominated by transporters. However, it was found that K^+ transporters and channels may operate in parallel in the plasma membrane of root cells (Garciadeblás et al., 2002). Transporters would have their range of activity at micromolar K^+ concentrations, whereas transport at millimolar concentrations would be mediated by K^+ channels (Rodríguez-Navarro and Rubio, 2006). In contrast to this notion, it is now evident that some channels mediate the transport of K^+ at micromolar concentrations (Dennison et al., 2001), that some HKT transporters are Na^+ transporters (Fairbairn et al., 2000; Uozumi et al., 2000; Horie et al., 2001), and that some KT-HAK transporters may mediate exclusively low affinity K^+ uptake (Senn et al., 2001). Taking into account the above, the main key issue to be addressed concerning K^+ homeostasis mechanisms in seagrasses is whether HAK transporters are only involved in high-affinity K^+ uptake, whereas channels carry out the uptake at millimolar K^+ concentrations.

Maintenance of appropriate intracellular K^+/Na^+ balance is critical for metabolic function as Na^+ cytotoxicity is largely due to competition with K^+ for binding sites in enzymes essential for cellular functions (Flowers and Colmer, 2008; Pardo, 2010; Kronzucker and Britto, 2011; Pardo and Rubio, 2011). Another adverse effect of Na^+ cytotoxicity is the production of ROS (reactive oxygen species), which then in turn affect cellular structure and metabolism negatively (Bartels and Sunkar, 2005). Plant cells are much more intolerant to Na^+ than animal cells due to their lack of significant systems for regulating their Na^+ content. In the Na^+-abundant marine environment where early life evolved, the use of K^+ as a major cation for maintaining the osmotic and electrical equilibrium of cells (Rodríguez-Navarro, 2000; Rodríguez-Navarro and Rubio, 2006) evolved in parallel with mechanisms of K^+ uptake and Na^+ exclusion. Recently,

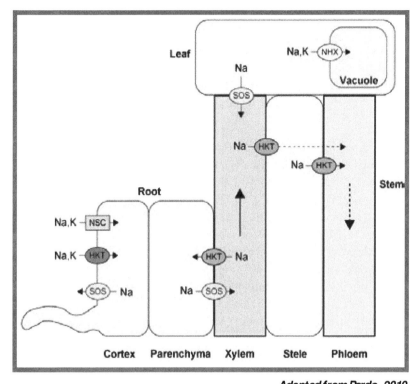

Adapted from Pardo, 2010

Figure 1. Schematic model for the function of SOS1, HKT proteins as well of nonselective cation channels (NSC) in achieving K⁺ uptake and Na⁺ exclusion in plants subjected to salinity stress.

it has been shown that a Na⁺-pump apparently does not exist in *Cymodocea nodosa*; on the contrary an electrogenic Na⁺/H⁺ antiporter seems to be the most likely mechanism that could mediate Na⁺ efflux in the epidermal cells of seagrasses (Apse and Blumwald, 2007; Garciadeblas et al., 2007; Touchette, 2007); however this activity has not yet been characterized. There is also evidence that H⁺-dependent systems are involved in the maintenance of a low cytosolic Na⁺ concentration in *Zostera marina* cells (Rubio et al., 2011).

The Na⁺/H⁺ antiporters in plants are electroneutral (Munns and Tester, 2008), which means they would not facilitate Na⁺ efflux at the alkaline pH values of seawater (Benito and Rodríguez-Navarro, 2003). However, seagrasses do presumably efflux Na⁺; their Na⁺/H⁺ antiporters function in this respect is unclear (Garciadeblás et al., 2007; Touchette, 2007; Flowers et al., 2010; Rubio et al., 2011). Recent molecular studies have demonstrated that genes encoding for Na⁺/H⁺ transporters (SOS1) are present in *C. nodosa* (Garciadeblás et al., 2007). SOS1 gene was initially cloned in the model plant *Arabidopsis thaliana*, encoding one of the plasma membrane Na⁺/H⁺ antiporter (Shi et al., 2000). Since then, SOS1-like genes have been detected in more than 30 terrestrial species, demonstrating its wide distribution in plants and its role in salinity tolerance as a sodium efflux mechanism (Rubio et al., 2011). The SOS1 system has been extensively investigated (Zhu, 2003) and it seems clear that it mediates Na⁺/H⁺ exchange (Shi et al., 2000) under the regulation of a protein kinase, SOS2, and a Ca^{2+}-binding protein, SOS3 (Qiu et al., 2002). Recent evidence indicates that the Na^+-induced stability of AtSOS1 mRNA is mediated by ROS (Chung et al., 2008). Presence of SOS1 system in *C. nodosa* suggests that this antiporter also play an important role in seagrass adaptation to the marine environment. Furthermore, this transport may show different characteristics than that of terrestrial plants, as suggested for salinity tolerance mechanisms in seagrasses (Touchette, 2007).

On the other hand, it has been proved that osmotic stress causes disorganization of microtubules in cells of higher plants (Yancey, 2001). Accumulation in the cytoplasm of non toxic compounds (osmolytes such as amino acids and methylamines) regulates osmotic homeostasis. The efficiency of osmolytes to act kosmotropically and not chaotropically, permit marine phanerogams to function under adverse conditions. In the plasmolysed cells peculiar tubular structures of microtubules are formed that appear to be related to the mechanism of regulation of protoplast volume. Moreover, actin cytoskeleton undergoes intense changes and thick bundles of actin microfilaments are formed (Komis et al., 2002a, b, 2003). A pivotal role to the cellular compartmental model of salinity tolerance response is the accumulation of metabolically 'compatible' organic solutes (osmolytes) in the cytoplasm, in order to balance the osmotic potential of Na^+ and Cl^- accumulated in the vacuole. Although, accumulation of osmolytes is required for osmotic cell homeostasis these compounds do not affect cellular functions (Jones and Gorham, 2002). Among the previously described osmolytes are amino acids such as proline, glycine, taurine, and methylamines such as betaine and trimethylamine N-oxide (TMAO; Touchette, 2007). Osmolytes appear to have additional functions, such as stabilizing proteins and membranes under conditions of dehydration, or by removing ROS. Osmoprotectant properties of compatible solutes include reduced inhibitory effects of ions on enzymes, increased thermal stability of enzymes, and limited dissociation of enzyme complexes (including the oxygen-evolving complex of photosystem II; Touchette, 2007). Little is known of the signaling cascades regulating the synthesis of osmolytes in seagrasses, although the molecular basis of NaCl-enhanced accumulation of some organic solutes has been studied in a few halophytes (Flowers and Colmer, 2008). During salinity stress, carbohydrates are likely converted to other organic compounds that would better facilitate osmotic adjustment in these plants. This is further supported by observed decreases in sucrose-P synthase (a key enzyme involved in sucrose synthesis) activities in seagrasses exposed to higher salinities (Touchette, 2007), where in *Ruppia maritima*, total soluble carbohydrate content appeared to decrease with increasing salinities (Murphy et al., 2003).

3. Methodology

Methodology should implement an innovative "cross-curricular" approach combining different interrelated scientific fields such as ecology, physiology, microscopy on cellular level, molecular biology/genetics and analytical biochemistry. This "cross-curricular" dimension reflects the capability of such an approach to incorporate successfully various scientific fields articulating its benefits to tackle the key issues in a functional, flexible and practical way. The selection and adoption policy should be based on the intention to support and to raise standards in marine ecology genetics research.

i.   Ecophysiology

Estimation of morphological and physiological parameters. Evaluation concerning the growth and photosynthetic (Fv/Fm και ΔF/Fm') response of seagrasses stress tolerance mechanisms on different levels of temperature, light intensity, PAR radiation and salinity. According to the literature review regarding seagrass species, it seems that the critical factors that affect their productivity and distribution in the Mediterranean Sea are temperature and PAR-radiation (Perez and Romero, 1992; Zharova et al., 2008). Particularly, at temperatures below 15°C and above 30°C flowering of species might be inhibited (i.e. Orfanidis et al., 2008; Sharon et al., 2009). Moreover, there is strong evidence to support the hypothesis that salinity, temperature and PAR-radiation fluctuations can critically affect seasonal distribution on a regional scale in certain phanerogam species (Gesti et al., 2005). Apart from the fact that seagrasses evolved by a common ancestor (high terrestrial plants) they seem to present relatively similar rapid growth rates with remarkable physiological plasticity, allowing them to respond and adapt to environmental stress, comprising them as ideal marine bio-indicators of environmental degradation.

ii.   Electronic microscopy on cellular level

The cellular structure (membranes, walls, organelles) mainly in the cytoskeleton organization and of the mechanism of the K^+/Na^+ pump function under various stress conditions using indirect fluorescent antibody (IFA) microscopy. The implementation of this technique allows successful spatial observation of cytoskeleton structures in cells. Otherwise, a Confocal Scanning Laser Microscope (CLSM) could be implemented. The main advantage of this method lies on the recombination of micro-slices in a three dimensional (3D) scale. The cellular mechanism of K^+-Na^+ pump function at different environmental stress conditions in means of plasma membrane vesicles would improve our knowledge on the adaptation mechanisms in terms of cell morphology.

iii.   Molecular biology/genetics

Isolation and characterization of HAK, SOS, HSPs and MT genes, which are putative gene-markers of the induced tolerance reactions under stress conditions. Whether any Na^+/H^+ antiporter activity is present at the plasma membrane of a leaf cell by in situ hybridization. In order to reveal homologous genes cloning of the corresponding cDNAs by using degenerated primers designed on highly conserved regions. Relative expression analysis by RT-qPCR evaluating the abundance of mRNAs in different tissues (indication of subcellular localization). The transcriptomic profiles (considering appropriate normalized libraries) in various abiotic stress conditions could be methodically determined. Enriched cDNA libraries could be thoroughly constructed by following the Illumina massively parallel sequencing technology (i.e. multiplex-based platform).

iv.   Analytical methods

Seagrasses under stressful conditions store in their vacuoles the toxic ions, such as Na^+. Therefore, the estimation of ions K^+, Na^+, Ca^{++}, Cl^- concentrations in different parts of the seagrass (root, rhizome, leafage, sheath) would contribute to the comprehension of their adaptive response mechanisms. Hence, the identification and quantitation of osmolytes by RP-HPLC with OPA derivatization could be justfully applied to illustrate the topic.

Due to the fact that the lot of terrestrial cultivated species do not tolerate high concentrations of NaCl, it would be beneficial to implement genetic improvement upon them in order to become tolerant in salinity. Thus, a new prospect would be the appraisal of economical exploitation by successful cultivation in high salinity soils.

4. Results and discussion

The primary effect of increased global temperature on seagrasses is the alteration of growth rates and other physiological functions of the plants themselves (Gaines and Denny, 1993; Gambi et al., 2009); it is also predicted that distribution of seagrasses will shift as a result of increased temperature stress in accordance with changes in the patterns of sexual reproduction (Short and Neckles, 1999; Gesti et al., 2005; Zharova et al., 2008). Identifying differentially expressed genes under stress is very useful in order to understand plant defense mechanisms (Rose et al., 2004; Whitehead and Crawford, 2006; Ouborg and Vriezen, 2007). Powerful techniques such as microarray analysis provide a wealth of information about genes involved in environmental stress responses and adaptation (Feder and Mitchell-Olds, 2003; Kore-eda et al., 2004; Ruggiero et al., 2004; Vasemägi and Primmer, 2005; Procaccini et al., 2007; Reusch and Wood, 2007). Many studies have shown up-regulation of transcripts for heat shock proteins (HSPs) (Rizhsky et al., 2002; Simões-Araújo et al., 2002; Busch et al., 2005; Huang and Xu, 2008; Larkindale and Vierling, 2008). Likewise, some studies have identified other transcripts increased by heat treatment, including members of the DREB2 family of transcription factors, AsEXP1 encoding an expansin protein, genes encoding for galactinol synthase and enzymes in the raffinose oligosaccharide pathway and antioxidant enzymes (Rizhsky et al. 2002, 2004; Busch et al. 2005; Lim et al. 2006; Xu et al. 2007). The most abundant transcript indentified was a putative metallothionein (MT) gene with unknown pleiotropic function, rich in cysteine residues in Z. marina (Reusch et al., 2008). In silico investigation in GenBank reveals the existence of orthologous gene counterparts coding for proteins with similar function (Bouck and Vision, 2007). The same appears to happen in the case of MTs (e.g. Guo et al., 2003) and in members of HSPs (e.g. Waters et al., 1996). Moreover, abundance and distribution of seagrasses are strongly related to the intensity of light. It comprises a significant factor for fitness, while it is also related to their photosynthetic capacity. The chlorophyll activity is considered as an adequate indicator of the biochemical and physiological robustness of plants (Vangronsveld et al., 1998); whilst salinity stress can alter photosynthetic capacities in seagrasses (Murphy et al., 2003; Torquemada et al., 2005). While increased salinity stress can cause declines in chlorophyll content (Baek et al., 2005; Karimi et al., 2005), other inhibitory processes are also involved including inhibition of electron flow, decreased photosystem function, diminished rubisco abundance and activity, and changes in chloroplast ultrastructure (Kirst, 1989; Ziska et al., 1990; Stoynova-Bakalova and Toncheva-Panova, 2003). The chlorophyll fluorescence is used in order to illustrate the stress degree due to an abiotic environmental factor, or combination of factors.

5. Conclusion

A comprehensive approach should consider the relative importance of each of the following components, thus providing a valuable insight on seagrasses multifunctional expression analyses.

i.   Marine angiosperms phenotype tolerance response from differential habitats to temperature, light and salinity fluctuations under controlled laboratory conditions.

    a.   Do factors such as space-time scales, environmental conditions, habitat type affect the variability of angiosperm species phenotype in representative coasts?

    b.   Estimation of morphological and physiological parameters. Measurements concerning the growth and photosynthetic (Fv/Fm, ΔF/Fm') response of seagrasess stress tolerance mechanisms on different levels of temperature, light intensity, PAR radiation and salinity.

ii.   Selective ion flux and ion portioning between cytoplasm and vacuole play an important role in establishing and maintaining different ion concentrations and ratios in seagrasses. However, the degree at which each mechanism is employed is not well understood. Exploration of the effects of various stress conditions on their cellular structure (membranes, walls, organelles) mainly in the cytoskeleton organization and of the mechanism of the K^+/Na^+ pump function with the implementation of electron microscopic techniques.

iii.   Comprehension of the molecular mechanisms involved at K^+ acquisition, Na^+ efflux and other pleiotropic responses.

iv.   Comparative genomic analysis of stress-specific cDNA libraries in order to evaluate the molecular homeostatic mechanisms that regulate tolerance reaction under various stress conditions.

    a.   Which are the genes of seagrasses that code for: 1) their HAK transporters, 2) their SOS antiporters that appear to intervene with Na^+ efflux, 3) their MT factors that pleiotropically intervene with the response at intense temperature fluctuations and 4) members of HSPs family that participate in the tolerance response induction under high temperatures.

    b.   Which is the transcriptomic profile for seagrass species that is induced in each stress factor?

v.   Identification and quantitation of osmolytes with osmoprotective activity. The estimation of ions K^+, Na^+, Ca^{++}, Cl^- concentrations in different parts of the seagrass (root, rhizome, leafage and sheath) will contribute to the comprehension of their adaptive response mechanisms.

vi.   Forecasting alterations in species distribution, abundance and diversity due to climate change.

vii.   Potential use of seagrass species as bioindicators of coastal and transitional waters.

Physiological, cellular, molecular and biochemical mechanisms which regulate stress tolerance responses in different levels of salinity, intense temperature fluctuations and light regime are not sufficiently studied in marine seagrasses. Our understanding of salinity tolerance in terrestrial halophytes and marine algae has considerably progressed over the last decade. Our knowledge of their variability according to species and habitat type is minimal. Nevertheless, several stress-related genes have been isolated and characterized in seagrasses. Such genes code protein transporters and antiporters which are related to the distribution of K^+ and Na^+ efflux, respectively, as well as genes coding for metallothionines (MT) and members of heat shock proteins (HSPs) family, which participate in pleiotropic response related to the intense temperature fluctuations and photosynthetic ability. Focusing studies to transcriptomic profiles and their equivalent metabolic pathways that regulate them, in combination with the assessment of the respective phenotype and the relevant physiological aspects, one can comprehend important ecological traits, such as tolerance in abiotic stress. As hectares of salt-affected land increases around the globe, understanding the origins of the diversity of seagrasses should provide a basis for the use of novel cultivated species in bioremediation and conservation.

Acknowledgements

This research has been co-financed by the European Union (European Social Fund-ESF) and Greek National funds through the Operational Program "Education and Lifelong Learning" of the National Strategic Reference Framework (NSRF)-Research Funding Program: Thales, Investing in knowledge society through the European Social Fund.

Author details

A. Exadactylos*

Address all correspondence to: exadact@uth.gr

Department of Ichthyology and Aquatic Environment, School of Agricultural Sciences, Univ. of Thessaly, Volos, Hellas, Greece

References

[1] Alberto, F., Massa, S., Manent, P., Diaz-Almela, E., Arnaud-Haond, S., Duarte, C.M. and Serrão, E.A. (2008). Genetic differentiation and secondary contact zone in the seagrass *Cymodocea nodosa* across the Mediterranean-Atlantic transition region. Journal of Biogeography, 35, 1279-1294.

[2] Apse, M.P. and Blumwald, E. (2007). Na$^+$transport in plants. FEBS Letters, 581, 2247-2254.

[3] Arnaud-Haond, S., Duarte, C.M., Alberto, F. and Serrão, E.A. (2007). Standardizing methods to address clonality in population studies. Molecular Ecology, 16, 5115-5139.

[4] Baek, M.H., Kim, J.H., Chung, B., Kim, J.S. and Lee, I. (2005). Alleviation of salt stress by low dose γ-irradiation in rice. Biologia Plantarum, 49, 273-276.

[5] Bartels, D. and Sunkar, R. (2005). Drought and salt tolerance in plants. Critical Reviews in Plant Sciences, 24, 23-58.

[6] Benito, B. and Rodríguez-Navarro, A. (2003). Molecular cloning and characterization of a sodium-pump ATPase of the moss *Physcomitrella patens*. The Plant Journal, 36, 382-389.

[7] Bloomfield, A. and Gillanders, B. (2005). Fish and invertebrate assemblages in seagrass, mangrove, saltmarsh, and nonvegetated habitats. Estuaries and Coasts, 28, 63-77.

[8] Bouck, A., and Vision, T. (2007). The molecular ecologist's guide to expressed sequence tags. Molecular Ecology 16 (5), 907-924.

[9] Burkholder, J.M., Tomasko, D.A. and Touchette, B.W. (2007). Seagrasses and eutrophication. Journal of Experimental Marine Biology and Ecology, 350, 46-72.

[10] Busch, W., Wunderlich, M. and Schoffl, F. (2005). Identification of novel heat shock factor-dependent genes and biochemical pathways in *Arabidopsis thaliana*. Plant Journal 41 (1), 1-14.

[11] Cabaço, S., Ferreira, Ó. and Santos, R. (2010). Population dynamics of the seagrass *Cymodocea nodosa* in Ria Formosa lagoon following inlet artificial relocation. Estuarine, Coastal and Shelf Science, 87, 510-516.

[12] Chung, J-S., Zhu, J-K., Bressan, R.A., Hasegawa, P.M. and Shi, H. (2008). Reactive oxygen species mediate Na$^+$-induced SOS1 mRNA stability in *Arabidopsis*. The Plant Journal, 53, 554-565.

[13] Costanza, R., d' Arge, R., de Groot, R., Farber, S., Grasso, M., Hannon, B., Limburg, K., Naeem, S., O'Neill, R.V., Paruelo, J., Raskin, R.G., Sutton, P., van den Belt, M.

(1997). The value of the world's ecosystem services and natural capital. Nature 387 (6630), 253-260.

[14] Directive 2000/60/EC (The EU Water Framework Directive) by: European Parliament. CELEX-EUR Official Journal L 327, 22 December 2000, p. 1-72.

[15] Directive 2008/56/EC (Marine Strategy Framework Directive) by: European Parliament and of the Council of 17 June 2008 establishing a framework for community action in the field of marine environmental policy, CELEX reference, 32008L0056, http://eur-lex.europa.eu/LexUriServ/LexUriServ.do?uri=OJ:L: 2008:164:0019:0040:EN:PDF

[16] Dennison, K.L., Robertson, W.R., Lewis, B.D., Hirsch, R.E., Sussman, M.R. and Spalding, E.P. (2001). Functions of AKT1 and AKT2 potassium channels determined by studies of single and double mutants of *Arabidopsis*. Plant Physiology, 127, 1012-1019.

[17] Duarte, C.M., and Prairie, Y.T. (2005). Prevalence of heterotrophy and atmospheric CO2 emissions from aquatic ecosystems. Ecosystems 8 (7), 862-870.

[18] Duffy, J.E. (2006). Biodiversity and the functioning of seagrass ecosystems. Marine Ecology Progress Series, 311, 233-250.

[19] Dupont, S., Wilson, K., Obst, M., Sköld, H., Nakano, H., and Thorndyke, M.C. (2007). Marine ecological genomics: When genomics meets marine ecology. Marine Ecology Progress Series 332, 257-273.

[20] Fairbairn, D.J., Liu, W., Schachtman, D.P., Gomez-Gallego, S., Day, S.R., and Teasdale, R.D. (2000). Characterization of two distinct HKT1-like potassium transporters from eucalyptus camaldulensis. Plant Molecular Biology 43 (4), 515-525.

[21] Feder, M.E. and Mitchell-Olds, T. (2003). Evolutionary and ecological functional genomics. Nature Reviews Genetics, 4, 649-655.

[22] Ferrat, L., Pergent-Martini, C. and Roméo, M. (2003). Assessment of the use of biomarkers in aquatic plants for the evaluation of environmental quality: application to seagrasses. Aquatic Toxicology, 65, 187-204.

[23] Flowers, T.J. (2004). Improving crop salt tolerance. Journal of Experimental Botany, 55, 307-319.

[24] Flowers, T.J. and Colmer, T.D. (2008). Salinity tolerance in halophytes. New Phytologist, 179: 945-963.

[25] Flowers, T.J, Galal, H.K. and Bromham, L. (2010). Evolution of halophytes: multiple origins of salt tolerance in land plants. Functional Plant Biology, 37, 604-612.

[26] Fu, H-H. and Luan, S. (1998). AtKUP1: A dual-affinity K+transporter from *Arabidopsis*. The Plant Cell Online, 10, 63-74.

[27] Fukuhara, T., Pak, J.Y., Ohwaki, Y., Tsujimura, H. and Nitta, T. (1996). Tissue-specific expression of the gene for a putative plasma membrane H⁺-ATPase in a seagrass. Plant Physiology, 110, 35-42.

[28] Gaines, S.D., and Denny, M.W. (1993). The largest, smallest, highest, lowest, longest, and shortest: Extremes in ecology. Ecology 74 (6), 1677-1692.

[29] Gambi, M.C., Barbieri, F. and Bianchi, C.N. (2009). New record of the alien seagrass *Halophila stipulacea* (Hydrocharitaceae) in the western Mediterranean: a further clue to changing Mediterranean Sea biogeography. Marine Biodiversity Records, 2, e84.

[30] Garciadeblas, B., Benito, B., and Rodríguez-Navarro, A. (2001). Plant cells express several stress calcium ATPases but apparently no sodium ATPase. Plant and Soil 235 (2), 181-192.

[31] Garciadeblás, B., Benito, B. and Rodríguez-Navarro, A. (2002). Molecular cloning and functional expression in bacteria of the potassium transporters CnHAK1 and CnHAK2 of the seagrass *Cymodocea nodosa*. Plant Molecular Biology, 50, 623-633.

[32] Garciadeblás, B., Haro, R. and Benito, B. (2007). Cloning of two SOS1 transporters from the seagrass *Cymodocea nodosa*. SOS1 transporters from *Cymodocea* and *Arabidopsis* mediate potassium uptake in bacteria. Plant Molecular Biology, 63, 479-490.

[33] Gesti, J., Badosa, A. and Quintana, X.D. (2005). Reproductive potential in *Ruppia cirrhosa* (Petagna) grande in response to water permanence. Aquatic Botany, 81, 191-198.

[34] Guo, W.J, Bundithya, W., and Goldsbrough, P.B. (2003). Characterization of the *Arabidopsis* metallothionein gene family: Tissue-specific expression and induction during senescence and in response to copper. New Phytologist 159 (2), 369-381.

[35] Hartog, den C. and Kuo, J. (2006). Taxonomy and biogeography in seagrasses. (In A.W.D. Larkum, R.R. Orth, & C.M. Duarte (Eds.), Seagrasses: Biology, Ecology and Conservation (pp. 1-23). The Netherlands: Springer).

[36] Heck Jr., K.L., Hays, G., and Orth, R.J. (2003). Critical evaluation of the nursery role hypothesis for seagrass meadows. Marine Ecology Progress Series 253, 123-136.

[37] Heck, K.L., Carruthers, T.J.B., Duarte, C.M., Hughes, A.R., Kendrick, G., Orth, R.J. and Williams, S.W. (2008). Trophic transfers from seagrass meadows subsidize diverse marine and terrestrial consumers. Ecosystems, 11, 1198-1210.

[38] Horie, T., Yoshida, K., Nakayama, H., Yamada, K., Oiki, S. and Shinmyo, A. (2001). Two types of HKT transporters with different properties of Na⁺and K⁺transport in *Oryza sativa*. The Plant Journal, 27, 129-138.

[39] Huang, B., and Xu, C. (2008). Identification and characterization of proteins associated with plant tolerance to heat stress. Journal of Integrative Plant Biology 50 (10), 1230-1237.

[40] Hughes, T.P., Baird, A.H., Bellwood, D.R., Card, M., Connolly, S.R., Folke, C., Grosberg, R., Hoegh-Guldberg, O., Jackson, J.B.C., Kleypas, J., Lough, J.M., Marshall, P., Nyström, M., Palumbi, S.R., Pandolfi, J.M., Rosen, B., Roughgarden, J. (2003). Climate change, human impacts, and the resilience of coral reefs. Science 301 (5635), 929-933.

[41] Ito, Y., Ohi-Toma, T., Murata, J. and Tanaka, N. (2010). Hybridization and polyploidy of an aquatic plant, *Ruppia* (Ruppiaceae), inferred from plastid and nuclear DNA phylogenies. American Journal of Botany, 97, 1156-1167.

[42] Jones, G. and Gorham, J. (2004). Intra-and inter-cellular compartmentation of ions. (In A.Läuchli, & U. Lüttge (Eds.), Salinity: Environment-Plants-Molecules (pp. 159-180). The Netherlands: Springer.)

[43] Karimi, G., Ghorbanli, M., Heidari, H., Khavarinejad, R.A. and Assareh, M.H. (2005). The effects of NaCl on growth, water relations, osmolytes and ion content in *Kochia prostrate*. Biologia Plantarum, 49, 301-304.

[44] Kim, E.J., Kwak, J.M., Uozumi, N. and Schroeder, J.I. (1998). AtKUP1: An *Arabidopsis* gene encoding high-affinity potassium transport activity. The Plant Cell Online, 10, 51-62.

[45] Kirst, G.O. (1989). Salinity tolerance of eukaryotic marine algae. Annual Review of Plant Physiology & Plant Molecular Biology 40, 21-53.

[46] Komis, G., Apostolakos, P. and Galatis, B. (2002). Hyperosmotic stress-induced actin filament reorganization in leaf cells of *Chlorophytum comosum*. Journal of Experimental Botany, 53, 1699-1710.

[47] Komis, G., Apostolakos, P. and Galatis, B. (2002). Hyperosmotic stress induces formation of tubulin macrotubules in root-tip cells of *Triticum turgidum*: Their probable involvement in protoplast volume control. Plant and Cell Physiology, 43, 911-922.

[48] Komis, G., Apostolakos, P. and Galatis, B. (2003). Actomyosin is involved in the plasmolytic cycle: gliding movement of the deplasmolyzing protoplast. Protoplasma, 221, 245-256.

[49] Kore-eda, S., Cushman, M.A., Akselrod, I., Bufford, D., Fredrickson, M., Clark, E., and Cushman, J.C. (2004). Transcript profiling of salinity stress responses by large-scale expressed sequence tag analysis in mesembryanthemum crystallinum. Gene 341 (1-2), 83-92.

[50] Kronzucker, H.J. and Britto, D.T. (2011). Sodium transport in plants: a critical review. New Phytologist, 189, 54-81.

[51] Larkindale, J. and Vierling, E. (2008). Core genome responses involved in acclimation to high temperature. Plant Physiology, 146, 748-761.

[52] Leoni, V., Vela, A., Pasqualini, V., Pergent-Martini, C. and Pergent G. (2008). Effects of experimental reduction of light and nutrient enrichments (N and P) on seagrasses: a review. Aquatic Conservation: Marine and Freshwater Ecosystems, 18, 202-220.

[53] Lim, U., Subar, A.F., Mouw, T., Hartge, P., Morton, L.M., Stolzenberg-Solomon, R., Campbell, D., Hollenbeck, A.R. and Schatzkin, A. (2006). Consumption of aspartame-containing beverages and incidence of hematopoietic and brain malignancies. Cancer Epidemiology, Biomarkers & Prevention, 15(9), 1654-9.

[54] Munns, R. and Tester, M. (2008). Mechanisms of Salinity Tolerance. Annual Review of Plant Biology, 59, 651-681.

[55] Murphy, L.R., Kinsey, S.T. and Durako, M.J. (2003). Physiological effects of short-term salinity changes on *Ruppia maritima*. Aquatic Botany, 75, 293-309.

[56] O'Neill, RV. (1988). Hierarchy theory and global change. (In T. Rosswall, R.G. Wood-mansee, & P.G. Risser (Eds), Scales and Global Change (pp. 29-45). NY: Wiley.)

[57] Orfanidis, S., Pinna, M., Sabetta, L., Stamatis, N. and Nakou, K. (2008). Variation of structural and functional metrics in macrophyte communities within two habitats of eastern Mediterranean coastal lagoons: natural vs. anthropogenic effects. Aquatic Conservation: Marine and Freshwater Ecosystems, 18, S45-S61.

[58] Orth, R.J., Caruthers, T.J.B., Dennison, W.C., Duarte, C.M., Fourqurean, J.W., Heck jr., K.I., Hughes, A.R., Kendrick, G.A., Kenworthy, W.J., Olyarnik, S., Short, F.T., Waycott, M. and Williams, S.I. (2006). A Global Crisis for Seagrass Ecosystems. BioScience, 56, 987-996.

[59] Ouborg, N.J. and Vriezen, W.H. (2007). An ecologist's guide to ecogenomics. Journal of Ecology, 95, 8-16.

[60] Pardo, J.M. (2010). Biotechnology of water and salinity stress tolerance. Current Opinion in Biotechnology, 21, 185-196.

[61] Pardo, J.M. and Rubio, F. (2011). Na$^+$and K$^+$transporters in plant signaling. (In: M. Geisler, & K. Venema (Eds.), Transporters and Pumps in Plant Signaling, vol. 7 (pp. 65-98). Berlin: Springer.)

[62] Pérez, M. and Romero, J. (1992). Photosynthetic response to light and temperature of the seagrass *Cymodocea nodosa* and the prediction of its seasonality. Aquatic Botany, 43 (1), 51-62.

[63] Procaccini, G., Olsen, J.L. and Reusch, T.B.H. (2007). Contribution of genetics and genomics to seagrass biology and conservation. Journal of Experimental Marine Biology and Ecology, 350, 234-259.

[64] Qiu, Q.S., Guo, Y., Dietrich, M.A., Schumaker, K.S. and Zhu, J.K. (2002). Regulation of SOS1, a plasma membrane Na$^+$/H$^+$exchanger in *Arabidopsis thaliana*, by SOS2 and SOS3. The Proceedings of the National Academy of Sciences USA, 99, 8436-8441.

[65] Quintero, F.J. and Blatt, M.R. (1997). A new family of K⁺transporters from *Arabidopsis* that are conserved across phyla. FEBS Letters, 415, 206-211.

[66] Reusch, T.B.H., and Wood, T.E. (2007). Molecular ecology of global change. Molecular Ecology 16 (19), 3973-3992.

[67] Reusch, T., Veron, A., Preuss, C., Weiner, J., Wissler, L., Beck, A., Klages, S., Kube, M., Reinhardt, R. and Bornberg-Bauer, E. (2008). Comparative analysis of expressed sequence tag (EST) libraries in the seagrass *Zostera marina* subjected to temperature stress. Marine Biotechnology, 10, 297-309.

[68] Rizhsky, L., Liang, H. and Mittler, R. (2002). The combined effect of drought stress and heat shock on gene expression in Tobacco. Plant Physiology, 130, 1143-1151.

[69] Rizhsky, L., Liang, H., Shuman, J., Shulaev, V., Davletova, S. and Mittler, R. (2004). When defense pathways collide: the response of *Arabidopsis* to a combination of drought and heat stress. Plant Physiology, 134, 1683-1696.

[70] Rodríguez-Navarro, A. (2000): Potassium transport in fungi and plants. Biochimica et Biophysica Acta (BBA)-Reviews on Biomembranes, 1469, 1-30.

[71] Rodríguez-Navarro, A. and Rubio, F. (2006). High-affinity potassium and sodium transport systems in plants. Journal of Experimental Botany, 57, 1149-1160.

[72] Rose, J.K.C., Bashir, S., Giovannoni, J.J., Jahn, M.M., and Saravanan, R.S. (2004). Tackling the plant proteome: Practical approaches, hurdles and experimental tools. Plant Journal 39 (5), 715-733.

[73] Rubio, F., Santa-María, G.E. and Rodríguez-Navarro, A. (2000). Cloning of *Arabidopsis* and barley cDNAs encoding HAK potassium transporters in root and shoot cells. Physiologia Plantarum, 109, 34-43.

[74] Rubio, L., Belver, A., Venema, K., Jesús García-Sánchez, M. and Fernández, J.A. (2011). Evidence for a sodium efflux mechanism in the leaf cells of the seagrass *Zostera marina* L. Journal of Experimental Marine Biology and Ecology, 402, 56-64.

[75] Ruggiero, M.V., Reusch, T.B.H. and Procaccini, G. (2004). Polymorphic microsatellite loci for the marine angiosperm *Cymodocea nodosa*. Molecular Ecology Notes, 4, 512-514.

[76] Santa-Maria, G.E., Rubio, F., Dubcovsky, J. and Rodriguez-Navarro, A. (1997). The HAK1 gene of Barley is a member of a large gene family and encodes a high-affinity potassium transporter. The Plant Cell Online, 9, 2281-2289.

[77] Schachtman, D.P. and Schroeder, J.I. (1994). Structure and transport mechanism of a high-affinity potassium uptake transporter from higher plants. Nature, 370, 655-658.

[78] Senn, M.E., Rubio, F., Bañuelos, M.A. and Rodríguez-Navarro, A. (2001). Comparative functional features of plant potassium HvHAK1 and HvHAK2 transporters. Journal of Biological Chemistry, 276, 44563-44569.

[79] Sharon, Y., Silva, J., Santos, R., Runcie, J.W., Chernihovsky, M. and Beer, S. (2009). Photosynthetic responses of *Halophila stipulacea* to a light gradient. II. Acclimations following transplantation. Aquatic Biology, 7(1-2), 153-157.

[80] Shi, H., Ishitani, M., Kim, C. and Zhu, J-K. (2000). The *Arabidopsis thaliana* salt tolerance gene SOS1 encodes a putative Na^+/H^+ antiporter. Proceedings of the National Academy of Sciences, 97, 6896-6901.

[81] Short, F.T. and Neckles, H.A. (1999). The effects of global climate change on seagrasses. Aquatic Botany, 63, 169-196.

[82] Simões-Araújo, J.L., Rodrigues, R.L., de A. Gerhardt, L.B., Mondego, J.M.C., Alves-Ferreira, M., Rumjanek, N.G. and Margis-Pinheiro, M. (2002). Identification of differentially expressed genes by cDNA-AFLP technique during heat stress in cowpea nodules. FEBS Letters, 515, 44-50.

[83] Stoynova-Bakalova, E. and Toncheva-Panova, T. (2003). Subcellular adaptation to salinity and irradiance in *Dunaliella salina*. Biologia Plantarum, 47, 233-236.

[84] Torre-Castro de la, M., and Rönnbäck, P. (2004). Links between humans and seagrasses-an example from tropical east Africa. Ocean and Coastal Management 47 (7-8), 361-387.

[85] Torquemada, Y., Durako, M. and Lizaso, J. (2005). Effects of salinity and possible interactions with temperature and pH on growth and photosynthesis of *Halophila johnsonii* Eiseman. Marine Biology, 148, 251-260.

[86] Touchette, B.W. (2007). Seagrass-salinity interactions: Physiological mechanisms used by submersed marine angiosperms for a life at sea. Journal of Experimental Marine Biology and Ecology, 350, 194-215.

[87] Touchette, B.W. and Burkholder, J.M. (2000). Review of nitrogen and phosphorus metabolism in seagrasses. Journal of Experimental Marine Biology and Ecology, 250, 133-167.

[88] Uozumi, N., Kim, E.J., Rubio, F., Yamaguchi, T., Muto, S., Tsuboi, A., Bakker, E.P., Nakamura, T. and Schroeder, J.I. (2000). The *Arabidopsis* HKT1 gene homolog mediates inward Na+currents in *Xenopus laevis* oocytes and Na+uptake in *Saccharomyces cerevisiae*. Plant Physiology, 122, 1249-1260.

[89] Vangronsveld, J.C.H.M., Cunningham, S.D., Lepp, N.W. and Mench, M. (1998). Pohysico-chemical aspects and efficiency of trace elements immobilization by soil amendments. In: Vangronsveld, J., Cunningham, S. D. (Eds), Metal-contaminated soils: In situ Inactivation and phytorestoration. R.G. Landes Co., Georgetown, TX, pp. 151-182.

[90] Vasemägi, A. and Primmer, C.R. (2005). Challenges for identifying functionally important genetic variation: the promise of combining complementary research strategies. Molecular Ecology, 14, 3623-3642.

[91] Vermaat, J.E., Verhagen, F.C.A. and Lindenburg, D. (2000). Contrasting responses in two populations of *Zostera noltii* Hornem. to experimental photoperiod manipulation at two salinities. Aquatic Botany, 67, 179-189.

[92] Vicente, O., Boscaiu, M., Naranjo, M.Á., Estrelles, E., Bellés, J.M. and Soriano, P. (2004). Responses to salt stress in the halophyte *Plantago crassifolia* (Plantaginaceae). Journal of Arid Environments, 58, 463-481.

[93] Waters, E.R., Lee, G.J., and Vierling, E. (1996). Evolution, structure and function of the small heat shock proteins in plants. Journal of Experimental Botany 47 (296), 325-338.

[94] Waycott, M., Procaccini, G., Les, D.H., and Reusch, T.B.H. (2006). Seagrass evolution, ecology and conservation: A genetic perspective. (In A.W.D. Larkum, R.R. Orth, & C.M. Duarte (Eds.), Seagrasses: Biology, Ecology and Conservation (pp. 25-50). The Netherlands: Springer.)

[95] Whitehead, A., and Crawford, D.L. (2006). Neutral and adaptive variation in gene expression. Proceedings of the National Academy of Sciences of the United States of America 103 (14), 5425-5430.

[96] Wiens, J.A., Stenseth, N.C., van Horne, B. and Ims, R.A. (1993). Ecological mechanisms and landscape ecology. Oikos, 66, 369-380

[97] Wissler, L., Codoner, F., Gu, J., Reusch, T., Olsen, J., Procaccini, G. and Bornberg-Bauer, E. (2011). Back to the sea twice: identifying candidate plant genes for molecular evolution to marine life. BMC Evolutionary Biology, 11, 8.

[98] Xu, I.Q., Sharp, D., Yuan, C.W., Yi, D.O., Liao, C.Y., Glaeser, A.M., Minor, A.M., Beeman, J.W., Ridgway, M.C., Kluth, P., Ager, J.W., Chrzan, III, D.C. and Haller, E.E. (2007). A Reply to the Comment by Frederic Caupin. Physical Review Letters, 99, 079602.

[99] Yancey, P.H. (2001). Water stress, osmolytes and proteins. American Zoologist, 41, 699-709.

[100] Zharova, N., Sfriso, A., Pavoni, B. and Voinov, A. (2008). Analysis of annual fluctuations of *Cymodocea nodosa* in the Venice lagoon: Modeling approach. Ecological Modelling, 216, 134-144.

[101] Zhu, J-K. (2003). Regulation of ion homeostasis under salt stress. Current Opinion in Plant Biology, 6, 441-445.

[102] Ziska, L.H., Drake B.G., and Chamberlain S. (1990). Long-term photosynthetic response in single leaves of a C3 and C4 salt marsh species grown at elevated atmospheric CO; *in situ*. Oecologia 83, 469-472.

3

Genetic Diversity and Conservation of an Endemic Taiwanese Species, *Platyeriocheir formosa*

Mei-Chen Tseng and Dai-Shion Hsiung

1. Introduction

Mitten crabs (Brachyura, Varunidae, Varuninae) are native to east Asia and currently classified into eight species belonging to four genera: *Eriocheir* De Haan, 1835, with *E. japonica* (De Haan, 1835) and *E. ogasawaraensis* Komai et al., 2006; *Paraeriocheir* gen. nov., with *P. hepuensis* (Dai, 1991) and *P. sinensis* (H. Milne Edwards, 1853); *Platyeriocheir* Ng et al., 1999, with *P. formosa* (Chan et al., 1995) and *P. guangdonga* sp. nov.; and *Neoeriocheir* T. Sakai, 1983, with *N. leptognatha* (Rathbun, 1913); plus an eighth, currently nameless species with the status of a *species inquirenda* [1].

Catadromous mitten crabs have the unusual life history of spawning in the sea and growing up in rivers. Fertilized eggs hatch into zoea, which leave the female and begin life in the sea as plankton. After passing through five ecdysis cycles over a period of several weeks, they metamorphose into megalopa (post-larval stage) that live in estuaries and migrate upstream to freshwaters where a second metamorphosis, into juvenile crabs, occurs [2]. The larval stage drifts passively with coastal currents, providing high potential for gene flow within coastal waters. Larvae mainly drift in proximity to the coastline rather than the open sea, so long distance dispersal across open seas is restricted [3]. Juvenile crabs move into rivers and dwell in their middle or upstream reaches where they grow until adulthood. They generally inhabit clear rushing waters as well as hiding in rock crevices by day and coming out at night to feed, their main food being periphyton growing on rocks and aquatic vegetation. They spend most of their lifetime (1–3 years) in freshwater and migrate downstream to coastal waters when mature to mate and spawn [3].

Of this group, only two genera and species, *Paraeriocheir hepuensis* and *Platyeriocheir formosa*, are native to Taiwan. The distribution of *P. hepuensis* in Taiwan extends from Dasi, Yilan

County to Wu Stream, Taichung City. *P. formosa*, a Taiwan endemic, is mainly distributed in rivers of eastern Taiwan (Figure 1). Both species are of economic importance; nevertheless, neither have been artificially cultured from egg to maturity, and as a result, people catch large numbers of them from the wild for sale. Therefore, these wild populations must be able to tolerate intense exploitation. In addition, they also suffer other adverse impacts, such as habitat destruction, natural population declines, climatic oscillations, and so on. Too, many rivers were severely impacted when typhoon Morakot hit the eastern and southern portions of Taiwan in 2009 by mudflows that covered riverbeds, resulting in a significant decrease in the *P. formosa* population size. In addition, wild crab catches are insufficient to meet consumer demand. Farmers directly import juvenile crabs of *P. sinensis* from China for culture until they are grown to adults for sale. However, when *P. sinensis* escape from farms and invade Taiwan rivers, hybrids with native species can occur to degrade the genetic structure of native species. At present, populations of these two native species are dwindling in population size due to habitat destruction by natural disasters and unrestricted over-harvesting. Moreover, *P. sinensis* has also been introduced into Taiwan for short-term aquaculture, disregarding the consequences of potential threats to native species. It is clear that conservation of *P. formosa* should be taken more seriously.

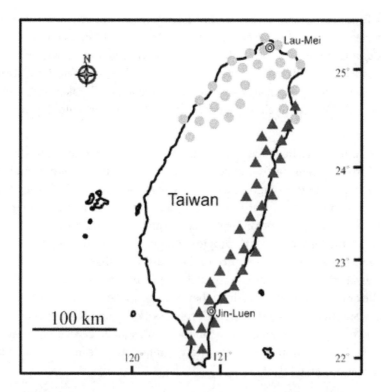

Figure 1. Map showing the distribution of two native mitten crabs in Taiwan and sampling sites. Black triangles indicated *Platyeriocheir formosa* and ashy circles indicated *Paraeriocheir hepuensis*.

P. formosa, *P. hepuensis*, and *P. sinensis* have similar morphological characters. Consequently, species identification was clarified using a molecular marker, mitochondrial DNA cytochrome oxidase subunit I (COI), which is often used in taxonomy, biodiversity assessments, phylogenetics, and phylogeographic studies [4-8]. Here we discriminate these three species using mt COI sequences. Because microsatellites have large mutation rates of $10^{-5} \sim 10^{-2}$ per generation [9-10], they are widely used as markers for studying genetics, population structure, kinship, and mating system [11-14]. In a previous study, the intraspecific genetic diversity of *P. formosa* was analyzed using COI sequences, with results showing insignificant genetic differences among samples from different streams [15]. To conserve *P. formosa* population diversity and ensure the sustainable use of this natural resource, its genetic diversity needed to be determined using polymorphic microsatellite loci.

This study attempted to distinguish *P. hepuensis*, *P. sinensis*, and *P. formosa*, and explored interspecies and intraspecific genetic diversity, using mt COI gene sequences. In addition, the genetic diversity of an endemic Taiwanese species, *P. formosa*, was examined by microsatellite loci. Effective management and conservative strategies also are proposed herein.

2. Experimental section

2.1. Sample collection

A total of 40 *Platyeriocheir formosa* specimens were collected in November 2010 from Jin-Luen, Taitung (120°55′ E, 22°32′ N), southeastern Taiwan. All *Paraeriocheir hepuensis* (n = 20) were collected from Lau-Mei Stream in New Taipei City and 2 *P. sinensis* individuals were collected from an aquaculture farm in Pingtung County.

2.2. Genomic DNA isolation

Muscle tissues from all specimens were preserved in 95% ethanol until DNA extraction. Genomic DNA was isolated and purified from the muscle tissue of all individuals. Five hundred milligrams of tissue with 1 mL lysis buffer was digested with 55 μL proteinase K solution. Small amounts of DNA were extracted for polymerase chain reactions (PCR) using a Puregene core kit A (Qiagen, Valencia, CA, USA).

2.3. COI subcloning and analysis

The complete COI gene was amplified using the specific forward primer 5′-CTCTAACR-GATTCCCCATCTTCTC-3′ and reverse primer 5′-ATCCTACACATCTGTCTGCC-3′ designed by the authors. A PCR consisted of approximately 50 ng genomic DNA, 50 pmol each of the forward and reverse primers, 25 mM dNTP, 0.05 ~ 0.1 mM $MgCl_2$, 10× buffer, and 5 U *Taq* polymerase (Takara Shuzo, Shiga, Japan), and brought up to 100 μL with Milli-Q water (Millipore, Billerica, MA, USA). The PCR program included one cycle of 4 min at 95 °C, 38 cycles of 1 min at 94 °C, 50 s at 50 °C, and 1 min at 72 °C, followed by a single further extension of 10 min at 72 °C. We evaluated 8 μL of each product on 0.8% agarose gel to check PCR success and confirm product sizes. The remaining PCR products were run on 0.8% agarose gels and purified using a DNA Clean/Extraction kit (GeneMark, Taichung, Taiwan). Purified DNA was subcloned into the pGEM-T easy vector (Promega, Madison, WI, USA) and transformed into *Escherichia coli* JM109. Plasmid DNA was isolated using a mini plasmid kit (Geneaid, Taichung, Taiwan). Clones from all individuals were sequenced on an Applied Biosystems (ABI, Foster City, CA, USA) automated DNA sequencer 377 (ver. 3.3) using a Bigdye sequencing kit (Perkin-Elmer, Wellesley, MA, USA).

In total, 42 COI sequences were subcloned. All sequences were aligned using Clustal W [16] and then checked with the naked eye. Intraspecific and interspecific genetic distances and numbers of different nucleotides were calculated using MEGA software [17]. The interspecific variable site numbers and intraspecific nucleotide diversities were computed by DnaSP v5 [18]. The phylogenetic trees for COI sequences were constructed using neighbor-joining [19] and maximum parsimony methods [20]. Cluster confidence was assessed using a bootstrap analysis with 1000 replications [21]. The minimum spanning tree (MST) was computed from the matrix of pairwise distances between all pairs of haplotypes in each sample using a modification of the algorithm [22]. We evaluated whether sequences had evolved under strict neutrality. Fu's *Fs* [23] and Tajima's neutrality tests [24] were performed in Arlequin 3.1 [25]. The significance of the statistics was tested by generating random samples under the hypothesis of selective neutrality and population equilibrium, using a coalescent simulation algorithm [26]. Tajima's test is based on an infinite-site model without recombinations. A significant D value can be due to factors other than selective effects, like population expansion, a bottleneck, or heterogeneity of mutation rates [27]. The possible occurrence of historical demographic expansions was examined using the mismatch distribution [28] implemented in Arlequin [25]. The distribution is unimodal in samples following a population demographic expansion [29].

2.4. Genotyping and data analysis

All 18 microsatellite loci [30] were amplified in this study. A PCR was performed in a volume of 25 μL that included ~10 ng genomic DNA, 10 pmol reverse primer, 10 pmol forward primer, 25 mM dNTP, 0.05–0.1 mM MgCl$_2$, 10× buffer, and 0.5 U *Taq* polymerase (Takara Shuzo, Tokyo, Japan) with Milli-Q water. The PCR products were subjected to a 1.5% agarose gel and allele sizes were checked by comparison with a DNA ladder and the length of the original sequence. Forward primers were labeled with FAM, TAMRA, or HEX fluorescence markers. PCR amplifications were carried out in a Px2 Thermal Cycler (Thermo Fisher Scientific, Waltham, MA, USA) with the following temperature profile: 1 cycle of 95 °C for 4 min, followed by 38 cycles of 94 °C for 30 s, and annealing at 50–60 °C for 30 s and 72 °C for 30 s. Each 5 μL of PCR product from three loci labeled with different fluorescence tags was mixed and precipitated with 95% alcohol. Semi-automated genotyping was performed using a capillary ABI 3730XL DNA Analyzer (ABI). Genotypes were scored with GeneMapper 4.0 (ABI).

The total number of alleles (*na*) and effective allele numbers were estimated for each locus using Popgene [31]. Observed (H_O) and expected (H_E) heterozygosities were independently calculated for each locus. Deviations from Hardy-Weinberg expectations (HWEs) were examined by an exact test using GENEPOP [32]. Linkage disequilibrium among all pairs of loci was determined using Burrow's composite measure [33] and χ2 values.

3. Results and discussion

3.1. Interspecific diversity

Molecular systematics and historical population dynamics were also analyzed using mt COI gene sequences. The full-length of all 42 COI sequences in *Platyeriocheir formosa* (n = 20), *Paraeriocheir hepuensis* (n = 20), and *P. sinensis* (n = 2) are consistent with 1534 bp in length. The average percentage of nucleotide components consist of 64% A + T for *P. sinensis*, 63% for *P. hepuensis*, and 62.6% for *P. formosa*, with slight differences among these three species.

The numbers of interspecific nucleotide differences ranged 196 - 221 with an average of 205.08 ± 4.14 between *P. formosa* and *P. hepuensis*, 199 - 214 with an average of 204.82 ± 3.30 between *P. formosa* and *P. sinensis*, and 60 - 79 with an average of 66.55 ± 3.87 between *P. hepuensis* and *P. sinensis*. Some interspecific nucleotide variable sites are shown in Figure 2.

The interspecific genetic distances ranged 0.142 - 0.163 with an average 0.150 ± 0.006 for *P. formosa* vs. *P. hepuensis*, 0.144 - 0.157 with an average 0.148 ± 0.002 for *P. formosa* vs. *P. sinensis*, and 0.041 - 0.054 with an average 0.045 ± 0.003 for *P. hepuensis* vs. *P. sinensis*. All 42 sequences in the study and one outgroup sequence from *Xenograpsus testudinatus* (NCBI accession number NC013480) were used to construct a phylogenetic tree by neighbor-joining (NJ) and maximum-parsimony (MP) methods. Phylogenetic trees presented significant clustering among the three species indicated *P. formosa*, *P. hepuensis*, and *P. sinensis* are an individually monophyletic group and share a common ancestor in two genealogical trees (Figure 3a, b). *P. hepuensis* and *P. sinensis* have closer relationships than *P. formosa*, indicating that the former two share a common recent ancestor. In addition, we also concluded that the COI gene is an effective genetic marker for distinguishing these mitten crabs having similar morphological characteristics.

The complete COI gene can be translated into a 511 amino acid sequence. The number of different amino acids ranged 0 - 10 within *P. formosa* containing two identical sequences and 0 - 16 within *P. hepuensis* containing four identical sequences. Intraspecific genetic distances ranged 0 - 0.020 (mean, 0.009 ± 0.004) within *P. formosa* and 0 - 0.033 (mean, 0.009 ± 0.007) within *P. hepuensis*. Interspecific genetic distance ranged 0.020 - 0.050 (mean = 0.026 ± 0.007) between *P. formosa* and *P. hepuensis*, 0.012 - 0.027 (mean = 0.019 ± 0.004) between *P. formosa* and *P. sinensis*, and 0.012 - 0.037 (mean = 0.017 ± 0.005) between *P. hepuensis* and *P. sinensis*. The NJ tree constructed from amino acid sequences reveals that all three species belong to one monophyletic group (Figure 4). The amino acid sequences of *P. hepuensis* and *P. sinensis* expressed higher similarity, suggesting that they have a closer relationship than either has with *P. formosa*.

```
[                1 1111111111 2222222222 2222222233 3333333333 3333334444 4444444444 ]
[       1225666792 2334456778 0012233344 5556778900 1156667777 8889990001 1122233344 ]
[       2125069203 9230718463 4702514536 2384392803 2843691258 1480692591 4706925847 ]
PF01 GCAGTCATTG CGTATATCAC AAAAATTTTT ATTTATTTTG TAGAAAGGCT TTCGGTTCGT GCGCGATTTT
PF02 .......... ...G....T. .......... .......... ...C.... .......... ........A. ..........
PF05 ..G....... ...G...... .........C ..C.C.... .......... .......... A......... ..........
PF06 .......... ...G...... .......... .......C... .......... .......... .......... ..........
PF07 .......... ...G...... .......... .......C... .......... .......... .......... ..........
PH01 A..A.T.CAA TACGCG...T G.CTG...C. GCAATGAGCA ATATT..ATC .ATAACCT.C AAATTTC.AC
PH02 A..A.T.CAA TA.GC....T G.CTG..... GCAATGA.CA ATATT..ATC .ATAACCT.C AAATTTC.AC
PH03 A..A.T.CAA TACGC....T G.CTG..CC. GCAATGA.CA ATATT..ATC .ATAACCT.C AAATTTC.AC
PH04 A..A.T.CAA TACGC....T G.CTG...C. GCAATGA.CA ATATT..ATC .ATAACCT.C AAATTTC.AC
PH05 A..A.T.CAA TACGC....T G.CTG...C. GCAATGA.CA ATATT..ATC .ATAACCT.C AAATTTC.AC
PS01 AT.ACTGCAA .ACG..AT.T .GCT.CC.C. .CAATAA.CA CTATTGA.TC CATAACCT.. AAATTTCCA.
PS02 AT.ACTGCAA .ACG..AT.T .GCT.CC.C. .CGATAA.CA CTATT...TC CATAACCT.. AAATTTCCA.

[       4444444444 4455555555 5555566666 6666666666 6666666777 7777777777 7777777788 ]
[       5556677888 8801123577 8889901122 3334556666 7788899001 1333444556 6788899900 ]
[       3692947023 6746981856 2384765814 0795170369 5814709581 4058147362 8103925817 ]
PF01 TGCTCGCGGC TACCCACTCG TCTCCCTATT CTAACTATAT TTCTTATTTC TTTACTCATC CGCGAACCAA
PF02 ........A. ........T. .......G.. .......... .......... .......... ..........
PF05 ........A. ........T. .......G.. .......... .......... .......... ..........
PF06 ........A. ........T. .......... .......... .......... .......... ..........
PF07 ........A. ........T. .......G.. .......... .......... .......... ..........
PH01 ..TCTATTAT CGTTTGTATA ATCAT.C.G. TCGGT.TA.A CATACGCCC. CC..TCT..T ATAT..TTGG
PH02 ..TCTATTAT C.TTTGTATA ATCAT.C.G. TCGGT.TA.A .ATACGC.C. CC..TCT..T ATAT..TTGG
PH03 ..TCTATTAT C.TTTGTATA ATCATTC.G. TCGGT.TA.A CATACGC.C. CC..TCT..T ATAT..TTGG
PH04 ..TCTATTAT C.TTTGTATA ATCAT.C.GC TCGGT.TA.A CATACGCCC. CC..TCT..T ATAT..TTGG
PH05 ..TCTATTAT C.TTTGTATA ATCAT.C.G. TCGGT.TA.A CATACGCCC. CC..TCTG.T ATAT..TTGG
PS01 CATCTATTAT C..TAGTATA CTCATAC.G. TC.GTC.AGA CATACGC.CT CCCGTC..CT ATATGGTT..
PS02 CATCTATTAT C..TAGTATA C.CATAC.G. TC.GTC.AGA CATACGC.CT CC.GTC..CT ATATG.TT..

[                1 1111111111 2222222222 2222222233 3333333333 3333334444 4444444444 ]
[       1225666792 2334456778 0012233344 5556778900 1156667777 8889990001 1122233344 ]
[       2125069203 9230718463 4702514536 2384392803 2843691258 1480692591 4706925847 ]
PF01 GCAGTCATTG CGTATATCAC AAAAATTTTT ATTTATTTTG TAGAAAGGCT TTCGGTTCGT GCGCGATTTT
PF02 .......... ...G....T. .......... .......... ...C.... .......... ........A. ..........
PF05 ..G....... ...G...... .........C ..C..C.... .......... .......... A......... ..........
PF06 .......... ...G...... .......... .......C... .......... .......... .......... ..........
PF07 .......... ...G...... .......... .......C... .......... .......... .......... ..........
PH01 A..A.T.CAA TACGCG...T G.CTG...C. GCAATGAGCA ATATT..ATC .ATAACCT.C AAATTTC.AC
PH02 A..A.T.CAA TA.GC....T G.CTG..... GCAATGA.CA ATATT..ATC .ATAACCT.C AAATTTC.AC
PH03 A..A.T.CAA TACGC....T G.CTG..CC. GCAATGA.CA ATATT..ATC .ATAACCT.C AAATTTC.AC
PH04 A..A.T.CAA TACGC....T G.CTG...C. GCAATGA.CA ATATT..ATC .ATAACCT.C AAATTTC.AC
PH05 A..A.T.CAA TACGC....T G.CTG...C. GCAATGA.CA ATATT..ATC .ATAACCT.C AAATTTC.AC
PS01 AT.ACTGCAA .ACG..AT.T .GCT.CC.C. .CAATAA.CA CTATTGA.TC CATAACCT.. AAATTTCCA.
PS02 AT.ACTGCAA .ACG..AT.T .GCT.CC.C. .CGATAA.CA CTATT...TC CATAACCT.. AAATTTCCA.

[       4444444444 4455555555 5555566666 6666666666 6666666777 7777777777 7777777788 ]
[       5556677888 8801123577 8889901122 3334556666 7788899001 1333444556 6788899900 ]
[       3692947023 6746981856 2384765814 0795170369 5814709581 4058147362 8103925817 ]
PF01 TGCTCGCGGC TACCCACTCG TCTCCCTATT CTAACTATAT TTCTTATTTC TTTACTCATC CGCGAACCAA
PF02 ........A. ........T. .......G.. .......... .......... .......... ..........
PF05 ........A. ........T. .......G.. .......... .......... .......... ..........
PF06 ........A. ........T. .......... .......... .......... .......... ..........
PF07 ........A. ........T. .......G.. .......... .......... .......... ..........
PH01 ..TCTATTAT CGTTTGTATA ATCAT.C.G. TCGGT.TA.A CATACGCCC. CC..TCT..T ATAT..TTGG
PH02 ..TCTATTAT C.TTTGTATA ATCAT.C.G. TCGGT.TA.A .ATACGC.C. CC..TCT..T ATAT..TTGG
PH03 .TCTATTAT C.TTTGTATA ATCATTC.G. TCGGT.TA.A CATACGC.C. CC..TCT..T ATAT..TTGG
PH04 ..TCTATTAT C.TTTGTATA ATCAT.C.GC TCGGT.TA.A CATACGCCC. CC..TCT..T ATAT..TTGG
PH05 ..TCTATTAT C.TTTGTATA ATCAT.C.G. TCGGT.TA.A CATACGCCC. CC..TCTG.T ATAT..TTGG
PS01 CATCTATTAT C..TAGTATA CTCATAC.G. TC.GTC.AGA CATACGC.CT CCCGTC..CT ATATGGTT..
PS02 CATCTATTAT C..TAGTATA C.CATAC.G. TC.GTC.AGA CATACGC.CT CC.GTC..CT ATATG.TT..
```

Figure 2. Partial interspecific variable sites within 12 COI sequences from *Platyeriocheir formosa*, *Paraeriocheir hepuensis* and *Paraeriocheir sinensis*.

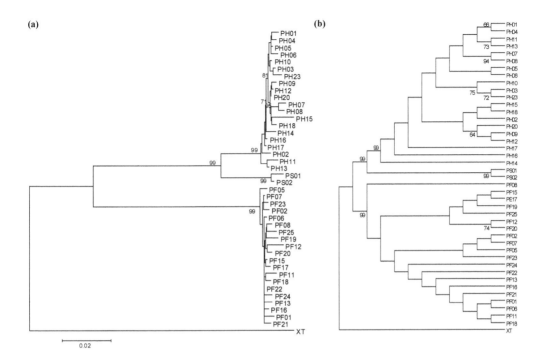

Figure 3. (a)Neighbor-joining tree and (b)Minimum parsimony tree constructed by 43 COI gene sequences from *Platyeriocheir formosa* (PF), *Paraeriocheir hepuensis* (PH) and *Paraeriocheir sinensis* (PS) and the outgroup *Xenograpsus testudinatus* (XT). Bootstrap values >60% (out of 1000 replicates) are shown at the nodes.

3.2. Intraspecific diversity and historical population dynamics

In total, 86 and 84 variable sites were respectively observed within intraspecific sequences of *P. formosa* and *P. hepuensis*. All 20 COI sequences from *P. formosa* contain seven highly variable sites (Figure 5a). The intraspecific number of nucleotide differences ranged 2 - 20. Haplotype diversity (H_d), the mean number of nucleotide differences (k), and mean nucleotide diversity (π) are 1, 9.73 ± 3.63, and 0.006 ± 0.003, respectively. A total of 86 substitutions containing 75 transitions and 11 transversions occur within these 20 sequences. There are 23 highly variable sites observed within 20 COI sequences of *P. hepuensis* (Figure 5b). Intraspecies nucleotide differences ranged 1 - 27. Haplotype diversity (H_d), the mean number of nucleotide differences (k), and mean nucleotide diversity (π) are 1, 11.36 ± 4.83, and 0.007 ± 0.004, respectively. The 85 substitutions include 68 transitions and 17 transversions. Intraspecific genetic distances of *P. formosa* ranged from 0.001 - 0.013, with an average of 0.006 ± 0.002. In contrast to *P. formosa*, *P. hepuensis* had similar intraspecific genetic distances that ranged from 0.001 to 0.018, with an average of 0.007 ± 0.003. Extremely high levels of H_d and low to moderate levels of π were discovered in these two species.

A similar genetic pattern is observed in many marine species [34]. The most likely explanation is that the accumulation of mutations over time in a rapidly growing population leads to an increase in the number of haplotypes; even so, population sizes suffer seriously when there is

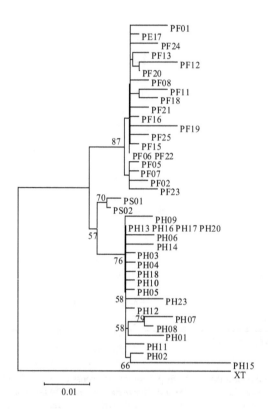

Figure 4. The neighbour-joining tree of COI amino acid sequences from *Platyeriocheir formosa* (PF), *Paraeriocheir hepuensis* (PH) and *Paraeriocheir sinensis* (PS) and the outgroup *Xenograpsus testudinatus* (XT). Numbers above the branches indicate the bootstrap values.

low genetic diversity. Except for wild mitten crabs, which must counteract overharvesting, probable natural causes include climate oscillations that result in temperature and water quality changes, mudflows covering riverbeds that can block migratory pathways, and any other environmental factor that affects adult reproduction and larval survival in estuaries. A second explanation for low levels of genetic diversity in these species could be due to their high dispersal potential during the planktonic egg and larval stages, resulting in strong gene flow among populations. Intraspecific genetic diversity in *P. formosa* analyzed by COI sequences indicate insignificant genetic differences among different populations [15]. It is interesting to note that *P. formosa* spawns in the sea and planktonic stage dispersal trends northward due to seasonal currents along the eastern Taiwan coast. This easily explains why *P. formosa* has low genetic variability among samples from different streams.

Four groups of marine fishes were defined based on the haplotype diversity (h_d) and nucleotide diversity (π) of various mtDNA coding regions [35]. The most widespread group possesses a high number of haplotypes ($H_d > 0.5$) and moderate to low levels of sequence divergences ($0.4\% < \pi < 0.8\%$). *P. formosa* and *P. hepuensis* were found in this study to have high haplotype diversity ($H_d = 1$) and moderate nucleotide diversities (0.6% & 0.7%), which fit the most common pattern

(a)

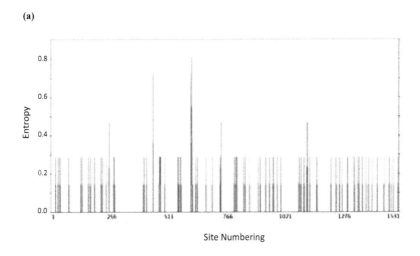

(b)

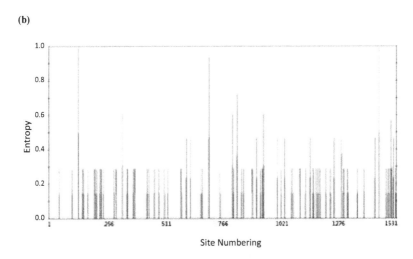

Figure 5. Seven and 23 highly variable sites observed within COI sequences. (a) *Platyeriocheir formosa* (b) *Paraeriocheir hepuensis*.

observed in marine fishes. High haplotype diversity within regional populations can be maintained through historically rapid population increases, resulting in the accumulation of mutations in populations [36]. Nevertheless, the shallow mtDNA branch structure of the NJ tree in these two mitten crabs might have resulted from catastrophic reductions in population size, which would produce low values of π.

The *D* values of Tajima's *D* neutral tests were analyzed to test this. *P. formosa* and *P. hepuensis* were -2.451 (p <0.001) and -2.129 (p <0.003), respectively. Negative Fu's *Fs* values of -12.646 (p = 0) in *P. formosa* and -11.333 (p = 0) in *P. hepuensis* suggested that these two species experienced a recent population expansion event. The mismatched distribution analysis presented average intraspecific nucleotide differences among COI sequences of 9.726 ± 7.731 in *P.*

formosa and 11.358 ± 6.344 in *P. hepuensis,* which were unimodally distributed (Figure 6) and indicated that the population experienced a historical expansion event. One and two central haplotypes were found in the minimum spanning tree (MST) of *P. formosa* and *P. hepuensis,* and most of the haplotypes were located at the tips (Figure 7), implying that adaptive radiation occurred.

(a)

(b)

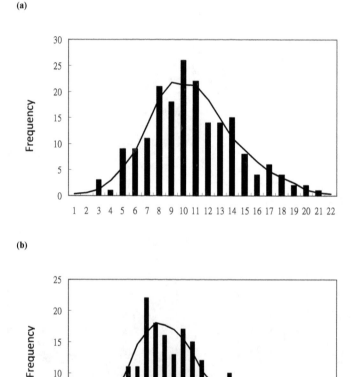

Figure 6. Mismatch distributions obtained from mtDNA COI data. The bars of the histogram represent the observed pairwise differences. The curve is the expected distribution under the sudden expansion model. (a) *Platyeriocheir formosa* (b) *Paraeriocheir hepuensis.*

Allele numbers and the effective allele numbers of all 18 microsatellite loci in *P. formosa* ranged from 3 - 14 and 2.25 - 10.26, respectively. Allele sizes within these loci ranged from 68 to 239 bp in length. The allele sizes of three loci (Pfo-15, -31, and -34) were all shorter than 100 bp (Table 1). Heterozygous individuals have been found at all loci except for Pfo-15. When

(a)   (b)

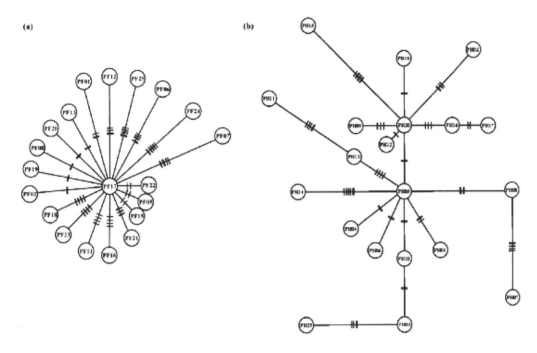

Figure 7. Minimum spanning tree constructed from COI data. (a) *Platyeriocheir formosa* (b) *Paraeriocheir hepuensis*.

excluding the homozygous Pfo-15 locus, the observed and expected heterozygosities (H_O & H_E) ranged from 0.20 - 0.95 (mean = 0.55) and 0.57 - 0.93 (mean = 0.818), respectively. Thirteen of these loci departed from the Hardy-Weinberg (HW) equilibrium, suggesting that *P. formosa* suffers from an intense inbreeding effect, bottleneck, or other possibility. Burrow's composite measure for linkage disequilibrium (LD) among the 18 loci was estimated for the entire dataset. The total variance of interlocus allele disequilibrium ($D_{IT}^2 = 0.018$) was slight. It is unlike the marine swimming crab, *Portunus trituberculatus*, which has a significantly higher mean heterozygosity ($H_O > 0.8$), while a similar H_O (0.55) was found in the catadromous *P. sinensis* [37]. The mean allele number per locus (*na*) in the catadromous *P. formosa* was estimated to be 9.61, which was higher than that of *P. sinensis* on average (*na* = 4.94), but lower than that of the marine *P. trituberculatus* (*na* = 22) [37-38]. These results agree with the observation that catadromous *P. formosa* have lower genetic diversity than marine *P. trituberculatus* species.

Microsatellite locus	Major repeats	T_a (°C)	Allelic size range (bp)	*na*	*ne*	H_O/H_E	NCBI accession no.
PFO-4	$(CA)_{29}$	56	145-183	13	6.45	0.75/0.87	JQ582816
PFO-5	$(TC)_6$	54	141-149	3	2.25	0.55/0.57	JQ582817
PFO-7	$(GT)_{22}$	52	217-239	10	6.06	0.65/0.86	JQ582818
PFO-9	$(CA)_{31}$	54	156-194	14	9.41	0.40*/0.92	JQ582819
PFO-10	$(CA)_{10}$	50	94-128	8	7.27	0.65/0.88	JQ582820
PFO-12	$(CA)_{32}$	56	186-216	14	7.84	0.45*/0.89	JQ582821

Microsatellite locus	Major repeats	T_a (°C)	Allelic size range (bp)	na	ne	H_O/H_E	NCBI accession no.
PFO-15	$(GT)_{16}$	60	68-76	5	2.99	0.00*/0.68	JQ582822
PFO-18	$(CA)_{33}$	60	134-170	13	6.84	0.35*/0.88	JQ582823
PFO-19	$(CA)_{32}$	50	141-175	13	9.20	0.25*/0.91	JQ582824
PFO-31	$(CA)_{17}$	52	79-93	6	2.74	0.35*/0.65	JQ582825
PFO-34	$(CA)_{20}$	50	71-77	3	2.38	0.70*/0.59	JQ582826
PFO-36	$(CA)_{35}$	54	101-125	10	6.11	0.50*/0.86	JQ582827
PFO-37	$(CA)_{31}$	50	166-200	13	8.79	0.90/0.91	JQ582828
PFO-51	$(CA)_{18}$	58	85-101	8	2.94	0.20*/0.68	JQ582829
PFO-52	$(CA)_{12}$	50	97-109	6	4.10	0.95*/0.78	JQ582830
PFO-54	$(GT)_{18}$	50	141-173	10	5.52	0.70*/0.84	JQ582831
PFO-60	$(GT)_{17}$	50	118-146	11	7.62	0.60*/0.89	JQ582832
PFO-79	$(GT)_{26}$	52	113-143	13	10.26	0.50*/0.93	JQ582833

Table 1. Characterization of the core region and levels of genetic variation at 18 microsatellite loci from *Platyeriocheir formosa*. T_a, PCR annealing temperature; *na*, observed number of alleles detected at each locus; *ne*, effective number of alleles; H_O, observed heterozygosity within a sample; H_E, expected heterozygosity within a sample. *significant Hardy-Weinberg deviation ($p < 0.05$).

P. formosa and *P. hepuensis* have a peculiar migratory history. Juvenile crabs migrate from the sea into rivers where they grow until adulthood. Mature adults move from their habitats in middle and upstream river reaches down to coastal waters for reproduction. Consequently, the most important conservation considerations are high water quality, freely flowing channels, reduced harvesting by humans, and preventing the invasion of the exotic *P. sinensis*. Focus must start on the catadromous journey of juvenile crabs from estuaries, and good water quality is the key to their survival and sustainable populations. Secondly, uninterrupted river flows are necessary, as mitten crabs must migrate throughout rivers in order to complete their full life cycle regardless of whether they are upstream-swimming juveniles or downstream-swimming adults. When river bottoms are buried under mudflows, previously established aquatic organisms are lost. However, temporarily created waterways will allow aquatic organisms to survive and complete their life cycles. Furthermore, overfishing results in decreasing crab resources and increases in their selling prices which results in additional overharvesting. However, capture can be banned for temporary periods of time to allow populations to recover. Finally, *P. sinensis* must be prevented from invading Taiwan's rivers. Because there are insufficient harvests of *P. formosa* and *P. hepuensis* to meet human demand, aquaculturists import large numbers of juvenile *P. sinensis* crabs and raise them to adulthood for sale. However, there are some problems with the culturing process. For one thing, *P. sinensis* effortlessly escapes from aquaculture even though ponds are equipped with anti-slipping nets. What is more, market prices may collapse if supply exceeds demand, and the industry might give up raising crabs and dump them. It is, therefore, certainly possible that these non-native crabs might colonize all of Taiwan's streams, causing an ecological catastrophe to native Taiwanese crab populations. Preventing the invasion of non-native mitten crabs should be of universal concern to Taiwan crab management and conservation.

The genetic structure and population dynamics of Taiwan's *P. formosa* population diversity must be continually monitored to ensure the sustainable use of this valuable natural resource.

4. Conclusions

The interspecific and intraspecific genetic diversity of two native Taiwanese mitten crabs, *P. formosa* and *P. hepuensis,* were determined in this study using the mtDNA COI gene and microsatellite loci. These two species possess similar genetic patterns with extremely high haplotype diversity and low to moderate nucleotide diversity. These results suggest that their population sizes historically underwent expansions but are currently undergoing serious decreases. Consequently, a conservation policy is proposed here that includes maintaining free-flowing stream channels and good water quality, preventing overharvesting by limiting harvesting to specific seasons, preventing the establishment of non-native mitten crabs, and conducting research on improved methods for the aquaculture of native mitten crabs.

Acknowledgements

We are extremely grateful to YH Hung for her help with laboratory work.

Author details

Mei-Chen Tseng* and Dai-Shion Hsiung

*Address all correspondence to: mctseng@mail.npust.edu.tw

Department of Aquaculture, National Pingtung University of Science and Technology, Taiwan

References

[1] Sakai K. A review of the genus *Eriocheir* De Haan, 1835 and related genera, with the description of a new genus and a new species (Brachyura, Grapsoidea, Varunidae). Crustaceana 2013;86:1103-38

[2] Shy JY, Yu HP. Complete larval development of the mitten crab *Eriocheir rectus* Stimpson, 1858 (Decapoda, Brachyura, Grapsidae) reared in the laboratory. Crustaceana 1992;63: 277-90

[3]   Xu J, Chan TY, Tsang LM, Chu KH. Phylogeography of the mitten crab *Eriocheir sensu stricto* in East Asia: Pleistocene isolation, population expansion and secondary contact. Molecular Phylogenetics and Evolution 2009;52: 45-56

[4]   Thaler AD, Plouviez S, Saleu W, Alei F, Jacobson A, Boyle EA, Schultz TF, Carlsson J, Dover CLV. Comparative population structure of two deep-sea hydrothermal-vent-associated decapods (*Chorocaris* sp. 2 and *Munidopsis lauensis*) from Southwestern Pacific Back-Arc Basins. PLoS One 2014; 9(7) e101345. doi: 10.1371/journal.pone.0101345

[5]   Zhan A, Bailey SA, Heath DD, Macisaac HJ. Performance comparison of genetic markers for high-throughput sequencing-based biodiversity assessment in complex communities. Molecular Ecology Resources 2014;14(5) 1049-59.

[6]   Korn M, Rabet N, Ghate HV, Marrone F, Hundsdoerfer AK. Molecular phylogeny of the Notostraca. Molecular Phylogenetics and Evolution 2013;69(3) 1159-71. doi: 10.1016/j.ympev.2013.08.006

[7]   Cornils A, Blanco-Bercial L. Phylogeny of the Paracalanidae Giesbrecht, 1888 (Crustacea: Copepoda: Calanoida). Molecular Phylogenetics and Evolution 2013;69(3) 861-72. doi: 10.1016/j.ympev.2013.06.018

[8]   Fernández MV, Heras S, Viñas J, Maltagliati F, Roldán MI. Multilocus comparative phylogeography of two aristeid shrimps of high commercial interest (*Aristeus antennatus* and *Aristaeomorpha foliacea*) reveals different responses to past environmental changes. PLoS One 2013;8(3) e59033. doi: 10.1371/journal.pone.0059033

[9]   Weber JL, Wong C. Mutation of human short tandem repeats. Human Molecular Genetics 1993;2: 1123-8

[10] Ellegren H. Mutation-rates at porcine microsatellite loci. Mammalian Genome 1995;6: 376-77

[11] Cristescu MEA, Colbourne JK, Radivojac J, Lynch M. A microsatellite-based genetic linkage map of the waterflea, *Daphnia pulex*: on the prospect of crustacean genomics. Genomics 2006;88: 415-30

[12] Kang JH, Park JY, Kim EM, Ko HS. Population genetic analysis and origin discrimination of snow crab (*Chionoecetes opilio*) using microsatellite markers. Molecular Biology Reports 2013;40(10) 5563-71

[13] Bentzen P, Olsen JB, McLean JE, Seamons TR, Quinn TP. Kinship analysis of Pacific salmon: insights into mating, homing, and timing of reproduction. Journal of Heredity 2001;92: 127-36

[14] Jensen PC, Bentzen P. A molecular dissection of the mating system of the Dungeness crab, *Metacarcinus magister* (Brachyura: Cancridae). Journal of Crustacean Biology 2012;32(3) 443-56

[15] Shiao CY. The genetic diversity of *Eriocheir formosa* (Crustacea, Decapoda, Grapsidae). Master Thesis. Department of Life Science, National Tsing Hua University, Hsinchu, Taiwan: 2007.

[16] Thompson JD, Higgins DG, Gibson TJ. CLUSTAL W: improving the sensitivity of progressive multiple sequence alignment through sequence weighting, position specific gap penalties and weight matrix choice. Nucleic Acids Research 1994;22(22) 4673-80. doi: 10.1093/nar/22.22.4673

[17] Tamura K, Stecher G, Peterson D, Filipski A, Kumar S. MEGA6: Molecular Evolutionary Genetics Analysis Version 6.0. Molecular Biology and Evolution 2013;30(12) 2725-2729. doi:10.1093/molbev/mst197

[18] Librado P, Rozas J. DnaSP v5: A software for comprehensive analysis of DNA polymorphism data. Bioinformatics 2009;25: 1451-2

[19] Saitou N, Nei M. The neighbor-joining method: A new method for reconstructing phylogenetic trees. Molecular Biology and Evolution 1987;4: 406-25

[20] Fitch WM. On the problem of discovering the most parsimonious tree. The American naturalist 1977;111: 223-57

[21] Hillis DM, Bull JJ. An empirical test of bootstrapping as a method for assessing confidence in phylogenetic analysis. Systematic Biology 1993;42(2) 182-92. doi: 10.1093/sysbio/42.2.182

[22] Rohlf FJ. Algorithm 76. Hierarchical clustering using the minimum spanning tree. Computer Journal 1973;16: 93-5

[23] Fu YX. Statistical tests of neutrality of mutations against population growth, hitchhiking and background selection. Genetics 1997; 147: 915-25

[24] Tajima F. Statistical method for testing the neutral mutation hypothesis by DNA polymorphism. Genetics 1989;123: 585-95

[25] Excoffier L, Laval LG, Schneider S. Arlequin vers. 3.0: an integrated software package for population genetics data analysis. Evolutionary Bioinformatics 2005;1: 47-50

[26] Hudson RR. Genealogies and the coalescent process. In: Furuyama D and Antonovics J (eds.). Oxford Surveys in Evolutionary Biology. New York: Oxford University; 1990

[27] Tajima F. The amount of DNA polymorphism maintained in a finite population when the neural mutation rate varies among sites. Genetics 1996;143: 1457-65

[28] Schneider S, Excoffier L. Estimation of past demographic parameters from the distribution of pairwise differences when the mutation rates vary among sites: application to human mitochondrial DNA. Genetics 1999;152: 1079-89

[29] Harpending H. Signature of ancient population growth in a low-resolution mitochondrial DNA mismatch distribution. Human Biology 1994;66: 591-600

[30] Cheng HL, Lee YH, Hsiung DS, Tseng MC. Screening of new microsatellite DNA markers from the genome of *Platyeriocheir formosa*. International Journal of Molecular Sciences 2012;13: 5598-606

[31] Yang RC, Yeh FC. Multilocus structure in *Pinus contoria* Dougl. Theoretical and Applied Genetics 1993; 87: 568-76

[32] Raymond M, Rousset F. GENEPOP (version 1.2) population genetic software for exact tests and ecumenicism. Journal of Heredity 1995;86: 248-9

[33] Cockerham CC, Weir BS. Quadratic analyses of reciprocal crosses. Biometrics 1977; 33: 187-203

[34] Bowen BW, Meylan AB, Ross JP, Limpus CJ, Balazs GH, Avise JC. Global population structure and natural history of the green turtle (*Chelonia mydas*) in terms of matriarchal phylogeny. Evolution 1992; 46: 865-81

[35] Grant WS, Bowen BW. Shallow population histories in deep evolutionary lineages of marine fishes: insights from sardines and anchovies and lessons for conservation. Journal of Heredity 1998;89: 415-26

[36] Rogers AR, Harpending H. Population growth makes waves in the distribution of pairwise genetic differences. Molecular Biology and Evolution 1992;9: 552-69

[37] Gao XG, Li HJ, Li YF, Sui LJ, Zhu B, Liang Y, Liu WD, He CB. Sixteen polymorphic simple sequence repeat markers from expressed sequence tags of the Chinese mitten crab *Eriocheir sinensis*. International Journal of Molecular Sciences 2010; 11: 3035-38

[38] Xu Q, Liu R. Development and characterization of microsatellite markers for genetic analysis of the swimming crab, *Portunus trituberculatus*. Biochemical Genetics 2011;49: 202-12

4

Advances in Genetic Diversity Analysis of *Phaseolus* in Mexico

Sanjuana Hernández-Delgado,
José S. Muruaga-Martínez,
Ma. Luisa Patricia Vargas-Vázquez,
Jairo Martínez-Mondragón, José L. Chávez-Servia,
Homar R. Gill-Langarica and
Netzahualcoyotl Mayek-Pérez

1. Introduction

Mexico is a centre of origin, domestication and diversity of major crops worldwide, such as maize (*Zea mays* L.) and beans (*Phaseolus* spp). It is known that five species of *Phaseolus* have been domesticated: *P. vulgaris* L. (common bean), *P. coccineus* L. ('ayocote' bean), *P. lunatus* L. ('lima' bean), *P. acutifolius* Gray ('tepary' bean) and *P. polyanthus* Greenm. (= *P. polyanthus* McFad.) ('acalete' bean) [1]. The five species are well distributed through Mexico as wild, semi-domesticated, and domesticated forms [2]. Beans are economically, socially, biologically and culturally important [3]. Owing to the importance of beans for Mexico and the world [4] extensive programmes focusing on the conservation, management and characterization of genetic resources of *Phaseolus* have been implemented in some countries [5] and international institutions such as Centro Internacional de Agricultura Tropical (CIAT) in Cali, Colombia; the Grupo de Mejoramiento de Leguminosas de la Misión Biológica de Galicia and Consejo Español de Investigación Científica (MBG-CSIC) in Spain; the USDA/ARS Western Regional Plant Introduction Station at USA, and the Instituto Nacional de Investigaciones Forestales, Agrícolas y Pecuarias (INIFAP) of Mexico who include bean collections in their germplasm banks, including all forms [6, 7].

Collections at germplasm banks are usually classified as base, active, for the work, and core [5]. Core collections include the highest levels of genetic diversity of one species (from 70 to 80%) as *Phaseolus* spp. into the lower number of accessions and due its size offers an easy and cheaper management and improves and efficient the germplasm use [8]. In this sense researchers need to characterize and to evaluate the core collections. Traditional strategies were based on the use of high-heritability morpho-agronomic traits such as growth habit, flower colour; the use of biochemical markers as seed reserve proteins as phaseolins, etc. Recent advances in molecular biology have improved the capability of plant genetic resource characterization using methodologies based on DNA analysis such as dominant (random amplified polymorphic DNA, or RAPDs, and amplified fragment length polymorphisms, or AFLPs) and co-dominant (restriction fragment length polymorphisms, or RFLPs, and simple sequence repeats, or SSRs) molecular markers [9]. The sequencing of the common bean genome has been concluded and this information broadens our perspective about some facts and challenges related to origin, domestication, diversity patterns and breeding of common beans [10-13].

2. Beans in Mexico

The common bean is the second legume crop worldwide, behind soybeans [*Glycine max* (L.) Merr] [4]. Despite the major anthropocentric use of beans being their green pods and dry grains, in some Latin American and African countries people consume young leaves and flowers as fresh vegetables [14]. By 2013, more than 1.8 million hectares were cultured with beans, and grain yields were 0.74 t ha^{-1} [15]. Common beans are well adapted, and grow through the different agroecosystems of Mexico and different seasons [7], depending upon the genetic diversity of the native germplasm and breed cultivars developed for each region [3].

Beans are a basic food due to being a major source of proteins, minerals, fibre, carbohydrates and vitamins in the Mexican daily diet for most people, but mainly for those with low economic resources [3]. Beans are the perfect complement to the Mexican diet based on maize as 'tortillas' in order to substitute animal protein for proper nutrition [4, 6]. Clinical studies showed that bean consumption prevents or improves both cholesterol and glucose levels in the blood [16]. However, the consumption of common beans has some problems, since they contain anti-nutritional compounds such as polyphenols (condensed tannins and anthocyanins), and inhibitors of proteases as trypsin, lectins and phytic acid [17, 18]. These compounds limit bean consumption and prevent them need breeding programmes and/or industry treatments such as cooking.

Economic conditions and customer preferences cause a variation in bean consumption *per cápita* between countries, and between regions within countries. In Mexico, black beans are preferred in the southern regions, while 'Flor de Mayo' and 'Flor de Junio' beans are preferred in the central western regions. Finally 'pinto', 'bayo' and 'azufrado' beans are mostly consumed in the northern regions. While Mexican bean *per capita* consumption is 11 kg per year, some African countries report more than 40 kg [18].

3. Cultivated species of *Phaseolus*

As has been listed before, only five *Phaseolus* species have been domesticated, and wild and cultivated forms are well known: *P. vulgaris* L., *P. coccineus, P. lunatus, P. acutifolius*, and *P. dumosus* from over 70 *Phaseolus* species, subspecies and varieties reported for Mexico (Tables 1 and 2). Most *Phaseolus* species are well distributed through Mexico with the exception of cold humid temperate and extremely hot arid tropic climates. The species *P. coccineus* and *P. leptostachyus* are distributed in the largest numbers of climatic types [19].

The cultivated form of *P. vulgaris* is grown across Mexico, but the highest producers are the states of Zacatecas, Durango, Sinaloa, San Luis Potosí, Guanajuato and Chiapas. Common beans represent 95% of national bean consumption. Common beans show an annual cycle and exhibit broad variability in growth habits, biological cycle duration, adaptation to different altitudes and soil conditions, as well as a large variation in seed colour and shapes. The last decades the common bean crop has migrated from central and southern regions to the north. Bean crop growth has also changed from common cultivation association with maize or pumpkin to monoculture using cultivars with determinate growth habits. Wild populations of *P. vulgaris* are mainly distributed through major mountain systems: Sierra Madre Occidental, Sierra Madre del Sur, Eje Neovolcánico, and Sierra Madre Oriental from 760 to 2,250 m above sea level. This species has a climbing growth habit, thin pods and small seeds where greyish-spotted seed coat colours are common, although there are other colours including black, 'bayo', yellow, brown, etc. Pods are dehiscent when mature, and seeds can be latent due to seed coat hardness. In this species it is common to find segregant populations derived from the spontaneous crossing between wild plants and cultivated forms, due to both types co-habiting or co-existing by variable lengths of time or seasons [7, 20-22].

Ayocote beans, *P. coccineus*, are known as 'patol' or 'patola'. They are a perennial species that develop tuber roots, as well as being a vigorous species, with climbing plants. Ayocote beans are commonly grown at the highlands of the central states of Puebla, Tlaxcala, Hidalgo and Mexico. *P. coccineus* is alogamous due to incompatibility problems, and landraces and wild plants show a great level of genetic variability in seed colours, sizes and shapes. The species shows short-width pods with a low number of seeds per pod; the seeds are large and have thick coats. The cultivated form has an impermeable seed coat. The wild form is distributed in temperate regions from the highlands and mountains: Sierra Madre Occidental, Eje Neovolcánico and Sierra Madre Oriental, between 1,800 and 3,000 m above sea level [23-26].

Lima beans, *P. lunatus*, are named as 'comba' at the Balsas River depression (western Mexico) or 'ibes' at the Península of Yucatán (southeastern Mexico). As ayocotes, lima beans are perennial and climbing species, vigorous and with a late biological cycle. Seeds of cultivated populations are variable in colour, shape and size. The species tolerates high temperatures, drought stress and some insect pests. Lima beans are frequently inter-cropped with maize, but most of their vegetative development and reproductive phase is completed after the maturity of maize when high levels of sun radiation are available. The wild form is perennial and climbing, and is distributed in the lower-coastal lands of Mexico near the Gulf of Mexico, Pacific Ocean and the Peninsula of Yucatán. Some wild populations have been found at the

Balsas River depression (states of Michoacán, Jalisco, Colima and Nayarit) at altitudes of up to 1,600 m above sea level (masl) [27, 28].

The poor surface cropped with *P. acutifolius* is located mainly in northern and northeastern Mexico at altitudes of up to 1,800 masl. Tepary bean is an annual species with some landraces with an undetermined growth habit for monoculture. Despite the cultivated form having larger seeds than the wild population, some varieties conserve grain traits similar to their ancestors. Wild populations are distributed at Sierra Madre Occidental from the northern states of Sonora and Chihuahua to the central-eastern state of Michoacán [27].

Finally, the less cultivated *Phaseolus* species, *P. dumosus* (= *P. polyanthus*), named 'acalete' shows vigorous climbing growth habits, and is commonly cultured in the state of Oaxaca, associated with maize. Acalete seeds are medium to large and are highly variable in colour, shape and size. The seed coat is thick, as in *P. coccineus,* but is permeable to water. The wild form is inferred to be dispersed through the state of Oaxaca [19, 27].

Section	Species	Mexican states of distribution
A. Acutifolii	*P. acutifolius* A. Gray var. *acutifolius*	Jalisco, Nayarit, Baja California, Chihuahua, Durango, Sinaloa, Sonora
	P. acutifolius A. Gray var. *latifolius*	Nayarit, Chihuahua, Coahuila, Durango, Sinaloa, Querétaro, Sonora
	P. acutifolius A. Gray var. *tenuifolius*	Colima, Jalisco, Baja California, Chihuahua, Coahuila, Colima, Durango, Jalisco, Sinaloa, Sonora.
	P. parvifolius Freytag	Baja California, Chiapas, Chihuahua, Durango, Guerrero, Jalisco, Michoacán, Nayarit, Oaxaca, Sonora
B. Phaseoli	*P. vulgaris* L.	Chiapas, Durango, Guanajuato, Guerrero, Jalisco, Mexico, Michoacán, Morelos, Nayarit, Oaxaca, Puebla, Querétaro, Sinaloa, Tamaulipas, Veracruz
	P. costaricensis	In Costa Rica and Panamá, Central America
	P. dumosus Macfady	Chiapas
	P. albescens McVaugh ex R. Ramírez & A. Delgado	Jalisco, Michoacán
C. Coccinei	*P. coccineus* L. subsp. *coccineus* (12 varieties: *coccineus, parvibracteolatus, griseus, lineatibracteolatus, tridentatus, splendens, strigillosus, semperbracteolatus, condensatus, pubescens, argenteus, zongolicensis).*	Higher regions with temperate or cold climatic conditions from Chiapas, Oaxaca, Guerreo, Morelos, Puebla, Veracruz, Tlaxcala, Edo de Mexico, Hidalgo, Guanajuato, Michoacán, Jalisco, Nayarit, Zacatecas, Durango, Nuevo León, Tamaulipas, Sinaloa,

Section	Species	Mexican states of distribution
	P. coccineus L. subsp. *striatus* (7 varieties: *striatus, minuticicatricatus, guatemalensis, purpurascens, rigidicaulis, pringlei, timilpanensis*)	Idem Subsp. *coccineus*
	P. glabelus Piper	Chiapas, Hidalgo, Oaxaca, Puebla, San Luis Potosí, Tamaulipas, Veracruz
D. Paniculati Sub-section I, Volubili.	*P. lunatus* L.	Baja California, Campeche, Chiapas, Colima, Guerrero, Jalisco, Mexico, Michoacán, Morelos, Nayarit, Oaxaca, Sinaloa, Tabasco, Tamaulipas, Veracruz, Yucatán
	P. polystachyus (three subspecies)	Only in the United States of America
	P. salicifolius Piper	Durango, Sinaloa, Sonora
	P. maculatifolius Freytag & Debouck	Nuevo León
	P. dasycarpus Freytag & Debouck*	Veracruz
	P. longiplacentifer Freytag	Veracruz
D. Paniculati Sub-section II, Lignosi	*P. jaliscanus* Piper	Jalisco, Nayarit, Sinaloa, Michoacán
	P. scrobiculatifolius Freytag	Michoacán
	P. nudosus Freytag & Debouck	Jalisco
	P. albinervus Freytag & Debouck	Chihuahua
	P. marechalii Delgado	Mexico, Morelos, Puebla
	P. rotundatus Freytag & Debouck	Jalisco, Michoacán
	P. acinaciformis Freytag & Debouck	Oaxaca
	P. xolocotzii Delgado	Guerrero, Mexico, Oaxaca
	P. sonorensis Standl	Chihuahua, Sinaloa, Sonora
	P. juquilensis Delgado	Oaxaca
E. Bracteati	*P. macrolepis* and *P. talamancensis*	Only in Guatemala and Costa Rica, Central America
F. Minkelersia	*P. pluriflorus* Maréchal, Mascherpa & Stainer	Distrito Federal, Durango, Mexico, Jalisco, Michoacán, Morelos, Nayarit, Sinaloa
	P. nelsonii Maréchal, Mascherpa & Stainer	Chiapas, Jalisco, Mexico, Michoacán, Oaxaca, Zacatecas
	P. perplexus Delgado	Jalisco, Mexico, Michoacán
	P. plagiosylix Harms	Nuevo León

Section	Species	Mexican states of distribution
	P. amblyosepalus (Piper) Morton	Durango, Michoacán, Sinaloa
	P. tenellus Piper	Mexico, Michoacán, Zacatecas
	P. parvulus Greene	Chihuahua, Durango, Nayarit, Sinaloa, Sonora, Zacatecas
	P. anisophylus (Piper) Freytag & Debouck	Chihuahua, Durango
	P. amabilis Standl	Chihuahua
	P. pausiflorus Sessé & Mociño	Chihuahua, Distrito Federal, Durango, Guerrero, Jalisco, Mexico, Michoacán, Morelos, Nayarit, Sinaloa, Sonora, Zacatecas
G. Zanthotricha	*P. xanthotrychus* Piper	Chiapas
	P. hintonii Delgado	Durango, Jalisco, Mexico, Nayarit
	P. zimapanensis Delgado	Durango, Hidalgo, Nuevo León, Querétaro, San Luis Potosí, Tamaulipas
	P. gladiolatus Freytag & Debouck	Hidalgo, San Luis Potosí
	P. magnilobatus Freytag & Debouck	Durango, Jalisco
	P. esquincensis Freytag	Chiapas
H. Revoluti	*P. leptophyllus* G. Don	Guerrero
I. Digitati	*P. neglectus* Hermann	Nuevo León, Tamaulipas
	P. albiflorus Freytag & Debouck	Coahuila, Nuevo León, Tamaulipas
	P. albiviolaceus Freytag & Debouck	Tamaulipas, Nuevo León
	P. trifidus Freytag	Nuevo León
	P. altimontanus Freytag & Debouck	Nuevo León
J. Rugosi	*P. filiformis* Benth.	Baja California, Chihuahua, Coahuila, Sonora
	P. angustíssimus A. Gray	Sonora
	P. carteri Freytag & Debouck	Baja California
	P. microcarpus Mart	Chiapas, Guanajuato, Durango, Guerrero, Jalisco, Michoacán, Oaxaca, Puebla
K. Falkati	*P. micranthus* Hook & Arn	Jalisco, Michoacán, Nayarit, Sinaloa, Sonora
	P. leptostachyus Benth. (five varieties: *leptostachyus, intosus, pinnatifolius, nanus, lobatifolius*)	Chiapas, Chihuahua, Colima, Distrito Federal, Durango, Guanajuato, Guerrero, Hidalgo, Jalisco, Mexico, Michoacán, Morelos, Nayarit, Nuevo León, Oaxaca, Puebla, Querétaro, San Luis Potosí, Sinaloa, Sonora, Tamaulipas, Veracruz, Zacatecas
	P. opacus Piper	Tamaulipas, Veracruz

Section	Species	Mexican states of distribution
	P. persistentus Freytag & Debouck	Guatemala
	P. macvaughii Delgado	Baja California, Colima, Guerrero, Jalisco, michoacán, Sinaloa
L. Brevilegumeni	*P. oligospermus* Piper	Chiapas
	P. campanulatus Freytag & Debouck	Nayarit, Jalisco
	P. tuerckheimii Donnell-Smith	Chiapas
M. Pedicellati	*P. pedicellatus* Benth	Distrito Federal, Guanajuato, Guerrero, Hidalgo, Mexico, Morelos, Nuevo León, Querétaro, San Luis Potosí, Tamaulipas, Veracruz, Jalisco, Michoacán
	P. oaxacanus Rose	Oaxaca
	P. esperanzae Seaton	Hidalgo, Mexico, Michoacán, Puebla
	P. polymorphus S. Wats (two varieties: *polymorphus, albus*)	Aguascalientes, Coahuila, Durango, Guanajuato, Jalisco, Nuevo León,
	P. palmeri Piper	Zacatecas
	P. purpusii Brandegee	San Luis Potosí
	P. grayanus Woot	Chihuahua, Coahuila, Duarngo, San Luis Potosí, Sonora, Zacatecas
	P. scrabellus Benth.	Durango, Sinaloa, Sonora
	P. teulensis Freytag	Durango, Zacatecas
	P. pyramidalis Freytag	Chihuahua
	P. laxiflorus Piper	Hidalgo, Mexico, Veracruz
N. Chiapasana	*P. chiapasanus* Piper	Chiapas, Oaxaca
O. Coriacei	*P. maculatus* Scheele subsp. *maculates*	Aguascalientes, Chihuahua, Coahuila, Durango, Guanajuato, Hidalgo, Puebla, Querétaro, San Luis Potosí, Sonora, Tlaxcala, Zacatecas
	P. maculatus Scheele subsp. *ritensis*	Chihuahua, Durango, Jalisco, Nayarit, Sinaloa, Sonora, Zacatecas
	P. venosus Piper	Aguascalientes, Durango, San Luis Potosí, Jalisco, Zacatecas
	P. reticulatus Freytag & Debouck	Durango

Adapted from Freytag and Debouck [29] and López-Soto et al. [19].

Table 1. Species of *Phaseolus* in Mexico and their distribution.

Taxon	Distribution	Altitudinal distribution (masl)
Section *Chiapasana*	Oaxaca, Veracruz, Chiapas	1000-1500
P. chiapasanus		
Section *Phaseolus*		
P. angustissimus	Chihuahua, Coahuila	1600-1750
P. filiformis	BCN, BCS, Sonora, Sinaloa, Chihuahua, Durango	200-1350
P. leptostachyus	Most of the country	500-2200
P. vulgaris	Most of the country	700-2000
P. acutifolius	BCS, BCN, Sonora, Sinaloa, Chihuahua, Durango, Zacatecas, Nayarit, Coahuila, Oaxaca, Guerrero, Morelos	20-2300
P. microcarpus	Jalisco, Oaxaca, Puebla, Veracruz, Guerrero, Morelos, Durango, Nayarit	20-1600
P. lunatus	Mexican coastal states, Puebla, Morelos	10-1200
P. neglectus	Tamaulipas, Nuevo León, Coahuila	1200-1600
P. coccineus	Medium to highlands of most of the country	700-2900
P. salicifolius	Durango, Sinaloa	1400-1900
P. maculatus	Chihuahua, Durango, Sonora, Zacatecas, Sinaloa, Aguascalientes, Guanajuato, Querétaro, Hidalgo, Jalisco, Nuevo León	1400-2200
P. polystachyus	Veracruz, Morelos	1200-1600
P. xolocotzii	Oaxaca, Morelos	1500-1800
P. ritensis	Chihuahua, Durango	1800-2100
P. marechalii	Hidalgo, Tlaxcala, Mexico	2100-2600
P. pedicellatus	Chihuahua, Durango, Aguascalientes, Zacatecas, SLP, Hidalgo, Oaxaca, Querétaro	1500-2800
P. sonorensis	BCS, BCN, Sonora, Chihuahua, Durango	40-1600
Section *Minkelersia*		
P. pauciclorus	Durango, Michoacán, Jalisco, Mexico, Oaxaca	1700-2000
P. nelsonii	Oaxaca, Mexico, Jalisco	1900-2200
P. parvulus	Chihuahua, Durango	2100-2500
P. pluriflorus	Mexico, Michoacán, Oaxaca, Jalisco	2000-2500
Section *Xanthotrichus*		
P. xanthotrichus	Hidalgo, Chiapas	1500-1700
P. hintonii	Oaxaca, Morelos, Michoacán	1500-1700

SLP = San Luis Potosí, BCN = Baja California Norte, BCS = Baja California Sur.

Table 2. Distribution of wild species of *Phaseolus* in Mexico based on specimens of INIFAP's herbarium.

4. Why must common bean diversity be conserved?

Mexico is broadly recognized as the centre of origin, domestication and diversification of major crops including avocado, amaranth, cocoa, pumpkin, maize, beans, and others. These species have been well dispersed and cultivated worldwide and constitute a major source of economic input for many countries. One major Mexican institution advocated to the collection, study, documentation, preservation and promotion of Mexican genetic resources is the Instituto Nacional de Investigaciones Forestales, Agrícolas y Pecuarias (INIFAP) which created the Genetic Resources Unit, and recently a new National Bank of Germplasm (CNRG, INIFAP) located in Tepatitlán, Jalisco, where *Phaseolus* represents one of the most important species to be preserved and studied. Special emphasis must be given in further works to those species not properly represented in Mexican Germplasm Banks, without seed samples preserved under proper conditions or species only represented as herbarium samples (Table 3). The recent explorations and collections of *Phaseolus* species across Mexico (Fig. 1) conducted by Muruaga-Martinez et al. (unpublished data, 2010-2012), showed the evident genetic erosion to be abnormally high in some regions. At these regions genetic erosion is so fast that can not will compensate by new genetic variability 'normally created' inside each ecosystem. Genetic erosion is significantly increased by human perturbation of agroecosystems. In addition, economical financing of a new re-collection of *Phaseolus* genetic resources must be revalued and consistently supported, assuming this involves future investment for genetic resource preservation and utilization for bean improvement.

Species	With seed available		Without seed available	
	Accessions registered in herbarium	Accessions with seeds	Species	Accessions registered in herbarium
P. acutifolius (three subspecies)	30	74	P. acinacifolius	1
P. amblyosepalus	5	2	P. albescens	2
P. coccineus (two subspecies)	123	282	P. albiflorus	8
P. chiapasanus	5	1	P. albinervus	1
P. esperanzae	12	1	P. albiviolaceus	2
P. glabellus	12	7	P. altimontanus	2
P. grayanus	25	8	P. amabilis	1
P. hintonii	5	1	P. anisophyllus	2
P. jaliscanus	9	4	P. angustissimus	2
P. leptostachyus (5 subspecies)	106	203	P. campanulatus	1
P. lunatus	43	91	P. carteri	4
P. maculatus/maculatus	25	28	P. dacicarpus	1
P. maculatus/ritensis	22	17	P. esquincensis	1

| Species | With seed available | | Without seed available | |
	Accessions registered in herbarium	Accessions with seeds	Species	Accessions registered in herbarium
P. marechalii	3	1	P. filiformis	33
P. micranthus	11	1	P. gladiolatus	2
P. microcarpus	31	7	P. laxiflorus	3
P. neglectus	3	7	P. juquilensis	1
P. nelsonii	12	4	P. leptophyllus	1
P. oaxacanus	5	3	P. longiplacentifer	1
P. oligospermus	6	10	P. maculatifolius	1
P. parvulus	14	4	P. macvaughii	10
P. pedicellatus	34	13	P. magnilobatus	3
P. pluriflorus	15	11	P. nudosus	1
P. polymorphus	19	2	P. opacus	1
P. scabrellus	3	1	P. palmeri	1
P. tuerckheimii	6	13	P. parvifolius	17
P. vulgaris	32	58	P. pauciflorus	30
P. xanthotrichus	2	8	P. perplexus	4
			P. plagyosilix	2
			P. purpusii	1
			P. pyramidalis	2
			P. rotundatus	1
			P. salicifolius	3
			P. scrobiculatifolius	2
			P. sonorensis	2
			P. tenellus	3
			P. teulensis	2
			P. trifidus	1
			P. venosus	3
			P. xolocotzii	3
			P. zimapanensis	12

Table 3. Species of *Phaseolus* geo-positioned, with herbarium registers and seed samples available and species without seed samples [29, 30].

In this sense, traditional strategies for germplasm analysis can be improved and better understood by the use of DNA-based strategies as molecular markers. Breeding programmes underutilize the genetic diversity available because of the necessity of pre-breeding exotic germplasm. The hybridization between wild and domesticated types of Phaseolus from the same gene pools offer greater potential for enhancing crop variation due the partial reproductive isolation between Andean and Mesoamerican domesticated gene pools. Former evaluations of wild and semi-wild *Phaseolus* accessions have shown resistance to insects and diseases [21,22, 31-33] and higher N, Fe, and Ca content in seeds, which could contribute to the improvement of nutritional grain quality and grain yield [34]. Acosta-Gallegos et al. [34] suggested Phaseolus pre-breeding based on the use of information on gene pool origins, domestication syndrome traits, molecular diversity, and mapping data of the wild forms; the indirect screening for biotic and abiotic stresses; and marker-assisted selection [35, 36].

Figure 1. Locations of exploration and collection of *Phaseolus* species in Mexico by Muruaga-Martínez et al. (unpublished data, 2010-2012).

5. Analysis of genetic diversity in domesticated *Phaseolus* species

5.1. Common beans

One of the pioneer works on the study of *Phaseolus* genetic diversity was conducted by Debouck et al. [37] who described the ecological adaptations and geographical locations of cultivated and wild species at northwestern South America (Colombia, Ecuador, Perú). Germplasm (12 wild and 36 cultivated accessions) collected from 1985 to 1990 was analysed on the basis of phaseolin and isozyme patterns. Wild beans showed a discontinuous distribution that, compared with bean distribution in Mexico, Central America and the Andean region, was classified as narrow. Later, Cerón et al. [38] analysed 151 bean accessions from the CORPOICA germplasm bank of Mosquera, Colombia. The data suggested higher morpho-agronomic variation in Mesoamerican germplasm than Andean accessions. Rodiño et al. [39] reported the analysis of 388 bean landraces from Spain and Portugal on the basis of the use of morpho-agronomic traits and phaseolin patterns. Germplasm was classified as Andean (74.7%), Mesoamerican (16.8%) and mixed (8.4%). The data indicated that only 52 accessions should constitute one representative core collection from Spain and Portugal.

In Mexico, the core collection of common beans structured by INIFAP includes 200 accessions, and this was characterized using morpho-agronomic traits and AFLP molecular markers. The data indicated a high level of genetic variability and no duplicity of accessions (non-shared haplotypes) into the core collection, becoming itself in a representative sample of *P. vulgaris* variation through Mexico [7, 40]. Rossi et al. [41] detected higher genetic diversity in the wild germplasm of beans from the Andean and Mesoamerican gene pools, while domesticated populations showed the largest linkage disequilibrium. Recent works of the analysis of *P. vulgaris'* genetic diversity have been published, where germplasm from different origins are analysed and broad genetic diversity ascertained into the genus and intra- and inter-specific and inter- and intra-population levels [42-48].

5.2. Ayocote beans

The genetic diversity of ayocote beans has been previously studied using morphological, agronomical and molecular markers, mainly using 'European' germplasm in order to detect high-yielding parents [49], highly tolerant germplasm to low temperatures [50], as well to characterize genetic relationships [51]. Spataro et al. [52] found clear differentiation between ayocote accessions [(wild, landraces, and *P. dumosus* (= *P. coccineus* ssp. *dumosus*)] as well as reduced gene diversity due the introduction of ayocote beans to Europe. Vargas-Vázquez et al. [24-26] reported that 80% of 798 *P. coccineus* accessions from Mexico originated from neo-volcanic Axis and Eastern Sierra Madre with humid or semi-arid temperate climates, from 1500-2000 m above sea level and at 500-1000 mm of annual precipitation. Ayocote bean germplasm can be separated into two groups: late accessions adapted to minimum temperatures (2-5°C), with large seeds and pods, and early accessions adapted to 0-2°C, with small seeds and pods. Analysis of European domesticated *P. coccineus,* including botanical varieties *albiflorus, bicolour* and *coccineus* and domesticated and wild accessions from Mesoamerica using cpSSRs, nuclear SSRs and phenotypic traits, suggested a moderate-to-strong cytoplasmic bottleneck that followed the expansion of species into Europe and multiple domestication events into the species. An adaptive population differentiation was also found, suggesting that selection led to the diversification of *P. coccineus* in Europe [53].

5.3. Lima beans

The genetic diversity of P. *lunatus* from the Yucatán Peninsula was assessed based on morphological and phenological characters, and then related to ethnobotanical information obtained about intraspecific diversity recognized by farmers, their selection criteria, agronomic management, production purpose and percentage of cultivated area. Ethnobotanical and morpho-phenological data indicated 30 putative distinct landraces, two wild, and two weedy variants from 149 seed samples of P. lunatus germplasm, suggesting gene flow among them. Richness and diversity estimates were greatest, and evenness lowest, where there was minimal agricultural intensification, wild and weedy populations, and greater persistence of traditional culture [28]. Afterwards, Martínez-Castillo et al. [54] determined genetic diversity, structure and gene flow of 11 wild populations of P. lunatus in four regions of traditional agriculture in the Yucatán Peninsula, Mexico, using SSR loci. Strong genetic differentiation was found among populations due to isolation among agricultural regions, as well as low long-term gene flow and low rates of recent migration among populations. Positive correlation between agricultural intensification and increased diversity was found because wild populations are favoured by the intensification of disturbance in situations involving at least three years of fallow.

Recent low gene flow at both intraregional and interregional levels into the wild-weedy-domesticated complex of P. *lunatus* under traditional agricultural conditions was found in four regions on the Yucatán Peninsula, Mexico, while gene flow from domesticated to wild populations was three times higher than in the opposite direction. This asymmetry was explained by regional agricultural practices and seed selection criteria. Domesticated alleles were shown to be entering wild populations of different agricultural regions, suggesting exchange of domesticated seeds between farmers of different regions. Thus, P. *lunatus* on the Yucatán Peninsula has a predominantly domesticated to wild gene flow, leading to genetic assimilation of the wild lima bean by its domesticated counterpart [55]. Afterwards, P. *lunatus* accessions collected in 1979 were compared with accessions collected in 2007 using SSR markers. The germplasm from 1979 was more diverse than that from 2007, suggesting the presence of a 'bottleneck' effect since alleles detected at each year of collection were different, as well as demonstrating allele drift due to the introduction of breed cultivars or changes in the selection criteria of germplasm [56].

Two wild Mesoamerican (MI and MII) gene pools with contrasting geographical distributions have been found in relation to P. lunatus. While the MI gene pool occurs in central western Mexico, including the Pacific coastal range, the MII gene pool is widespread and occurs towards the Gulf of Mexico, the Yucatán Peninsula, and Central and South America. Mesoamerican landraces clustered together with wild accessions from the MI gene pool (L haplotype), suggesting a unique domestication event in central western Mexico. The most likely domestication region is an area of the states of Nayarit–Jalisco or Guerrero–Oaxaca, and not areas such as the Peninsula of Yucatán where the crop is currently widespread and diverse. A strong founder effect due to domestication has been detected, and several recently diversified haplotypes identified [57]. The analysis of 67 wild populations of P. lunatus from Mexico with ten microsatellite loci confirmed not only the presence of the two gene pools (MI and MII), but also the possible existence of two subgroups within MI (MIa and MIb). While MI and MII are mainly divergent geographically, MIa and MIb overlap in their distribution. Thus, the genetic structure of the wild lima bean in Mexico is more complex than previously thought, and the presence of three gene pools (MIa, MIb, and MII), each one possessing relatively high levels of genetic diversity, is proposed [58]. Other work, including P. lunatus populations from different areas of America and germplasm, was analysed using two intergenic spacers of chloroplast DNA: atpB-rbcL and trnL-*trnF*. Three groups (AI, MI, MII) of genotypes were found, confirming the existence of Mesoamerican and Andean gene pools and multiple origins of domestication for the MA gene pool. For MI, western central Mexico was proposed as the domestication area, and for MII this was between Guatemala and Costa Rica [59].

5.4. Tepary beans

Since few genetic tools have been developed or tested for tepary bean, Blair et al. [60] validated one set of gene-derived and non-gene simple sequence repeat or microsatellite markers from the common bean in tepary bean cultivars and wild relative accessions. They then evaluated the genetic diversity and population structure of tepary bean accessions to determine if leaf morphology variants are valid as separate subgroups of wild tepary beans; if *P. parvifolius* was a separate variant or species; and if cultivated tepary beans originated from one domestication event or several events. The analysis of 140 tepary bean genotypes showed that a single domestication was likely as the cultivars were most closely related to accessions from Sinaloa and northern Mexico, and that diversity was much higher in the wild genotypes compared to the cultivated ones. *P. parvifolius* was classified as a separate species by population structure analysis while the variants *P. acutifolius* var. acutifolius and var. tenuifolius were admixed and inter-crossed. *P. latifolius* was not a valid species or variant of *P. acutifolius*, but represents a group of cultivars within the tepary bean. Other recent work was focused on the analysis of the agro-morphological variation of *P. acutifolius* germplasm in Botswana, where low genetic diversity was found [61].

5.5. Acalete beans

Total seed protein variability in a sample of 163 entries of year bean (*P. polyanthus*), including wild, feral and cultivated forms of the whole range of distribution in Latin America, was studied using I-dimensional SDS/PAGE and 2-dimensional IEF-SDS/PAGE. Ten different patterns were observed in this crop. Eight of these are found in the Mesoamerican materials, the other two of those in the northern Andes. The highest diversity is found in the wild ancestral forms present in central Guatemala with six patterns. The 'b' pattern predominant in all Mesoamerican cultivated materials and is also present at low frequency in Colombia. The 'k' pattern, predominant in the northern Andes, is present in Costa Rica. These results, together with information on indigenous names for the crop, suggest that there is a single gene pool domesticated from a wild ancestor still present in Guatemala, and distributed afterwards to the northern Andes, but with a clinal genetic drift from Mesoamerica to the Andean region [62].

6. Comparisons of genetic diversity among bean species

The diversity and relationships among species of *Phaseolus* complex were analysed using chloroplast DNA. Restriction patterns were used to identify polymorphisms and assess the type of mutations detected, and identify regions of high variability (Fig. 2). There is high cpDNA variability within *P. coccineus* but other species as *P. vulgaris* and *P. coccineus* subsp. *glabellus* show a very distinct cpDNA genotype compared to the former species. These evidences strongly suggests that *P. coccineus* subsp. *glabellus* belongs to a different but as yet undetermined section of the genus. In *P. coccineus* subsp. *darwinianus* (= *P. polyanthus),* the cpDNA lineage was in disagreement with data obtained from nuclear markers, and suggested a reticulated origin by hybridization between *P. coccineus* as the male parent and an ancestral *P. polyanthus* type, closely allied to *P. vulgaris.* Molecular markers are an important strategy for elucidating phylogenetic relationships; in addition, accurate phylogenies will require analyses of both nuclear and cytoplasmic genomes [63].

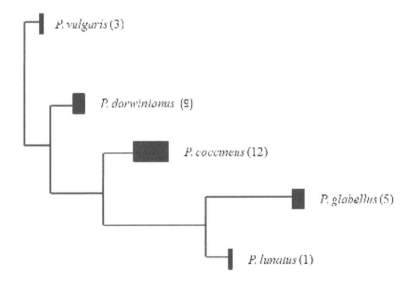

Figure 2. Dendrogram of five *Phaseolus* species analysed by cpDNA restriction patterns. Adapted from data of Llaca et al. [63].

Hamann et al. [64] identified 18 species from 90 genotypes using SSR markers, where the species *P. vulgaris, P. lunatus, P. coccineus, P. acutifolius* and *P. polyanthus* showed four specific (GATA) sequence patterns that help to clearly separate auto-pollinated (*P. vulgaris* and *P. lunatus*) from alogamous species as *P. coccineus,* but this species was found to have lower intra-specific variation. Gaitán-Solís et al. [65] later isolated, cloned and sequenced genomic DNA fragments into three gene libraries and then evaluated the polymorphisms of 68 SSRs. Markers were capable of separating germplasm on the basis of *Phaseolus* species: *P. coccineus, P. polyanthus, P. acutifolius,* and *P. lunatus.* Blair et al. [66] used genic and genomic microsatellites to analyse allele diversity and heterozygosity in *P. vulgaris* and *P. acutifolius*. Genic sequence SSRs were more polymorphic than genomic SSR. SSR distinguished between Mesoamerican and Andean gene pools, and separated genetic races into each gene pool and into wild from cultivated germplasm. Andean germplasm was more polymorphic at both inter- and intra-populational levels. Contrasting results were found by Benchimol et al. [67], who analysed Mesoamerican and Andean *P. vulgaris* accessions and reported low values of polymorphisms, likely due to the domestication process [68].

In Italy, 66 genotypes representing 14 local varieties of *P. vulgaris* and nine of *P. coccineus*, collected through regions of Marche using ISSR markers, SSRs and cpSSRs, were analysed. Farmers' selection and adaptability to variable environments have provoked bean preservation and diversification. A total of 71% of local varieties of *P. vulgaris* come from Andean origins [69]. Chacón et al. [70] analysed 31 accessions of *Phaseolus* (27 from *P. coccineus* including both *P. coccineus* subspecies *darwinianus* and *glabellus*; three from *P. vulgaris* and one from *P. lunatus*) using restriction patterns of cpDNA. Molecular analysis clearly differentiated between *Phaseolus* species, and *P. coccineus* showed the highest molecular polymorphism values in both wild and cultivated accessions compared with all other species [71, 72]. In Mexico, Ramírez et al. [73] characterized 107 common bean populations, 42 ayocote beans and one acalete bean on the basis of morpho-agronomic traits. Ayocote beans showed more diverse seed coat colours than common beans (54.8% purple, 26.2% black, 19.0% white and brown).

7. Our modest advances

7.1. About the origin, domestication and classification of *Phaseolus*

Hernández-López et al. [74] published a review paper that analysed classic works focused on determining and locating the centres of origin and domestication of *P. vulgaris*, assuming that these areas are major sources of populations carrying useful genes for breeding, and because such populations can improve our understanding of the evolution, diversification and conservation of the species. Despite the broad and abundant information published over decades, new information is consistently published, new evidence found, and new strategies such as genomic and genetic techniques based on DNA analysis applied in these studies. The accumulated knowledge derived from varied sources including archaeology, agronomy, ethno-linguistic, ethnobotany, molecular biology, biochemistry, physics etc. is currently being applied in order to study and clarify origins, domestication and diversification patterns, phylo-geographical relations, among others. Therefore, the use of tools based on molecular technologies and genomics should give definitive evidence on the origin, domestication and genetic diversity of *Phaseolus* [11, 74, 75].

As has been described by Muruaga-Martínez et al. (unpublished data), recent re-collection tours have been conducted in order to clarify the real and current state of genetic resources of *Phaseolus* across Mexico. One major problem for taxonomy, phylogeny or systematics studies in *Phaseolus* is that most specimens belong to herbariums and no 'fresh' plants are available. Thus, our work group assumed the necessity to re-collect *Phaseolus* specimens. We then

subjected the germplasm to genetic analysis using molecular marker strategies and repro-
duced them under controlled conditions for future works, and to preserve endangered species
and specimens. The first expeditions (2010-2012) yielded the collection of more than 100
samples (seeds), which comprised 19 species, including two subspecies of *P. coccineus* (*P.
coccineus griseus* and *P. coccineus striatus*). Three species in this collection (*P. albiviolaceus, P.
maculatifolius* and *P. rotundatus*) had not been studied before. Villarreal-Villagrán et al. [76]
analysed them by using five *trnT-trnL, trnL-trnF, rpl16, rpoC1-rpoC2, rps14-psaB* non-coding
regions of chloroplastic DNA amplified by PCR (polymerase chain reaction). Cluster analysis
confirmed with strong bootstrap support that the genus *Phaseolus* is a monophyletic group that
can subdivide itself into two major lineages: one includes *P. pluriflorus, P. esperanzae, P.
pedicellatus, P. microcarpus, P. glabellus, P. oligospermus, P. gladiolatus, P. zimapanensis* and *P.
albiviolaceus*; and the other includes *P. filiformis, P. acutifolius, P. vulgaris, P. coccineus striatus,
P. coccineus griseus, P. macvaughii, P. leptostachyus, P. lunatus, P. maculatus, P. maculatifolius* and
P. rotundatus. The topology of the dendrogram obtained agreed with the topology of *Phaseolus* recognized to this date, which was obtained using only the ribosomal ITS and chloroplast
trnK locus [77]. The exception was *P. albiviolaceus*, a species not studied before, that according
to traditional morphological criteria, belongs to the Pedicellatus group, but which in this study
appeared with the Tuerckheimii group. The other two species that were characterized for the
first time in a molecular phylogeny are *P. maculatifolius* and *P. rotundatus*, both of which were
clustered within the polystachios group (Fig. 3).

7.2. About the genetic diversity analyses of *Phaseolus*

The analysis of Mexican common bean core collection using SSR and AFLP markers revealed
that the highest genetic diversity is found in central Mexico and Chiapas, which seems to be
an important diversity centre in the south. SSR analysis indicated a reduced number of shared
haplotypes among accessions and core collection has no duplicated accessions [40]. Hernández-López et al. [21, 22] evaluated the diversity and genetic relations of one collection of bean
populations produced after the random crossing among wild and domesticated or cultivated
bean genotypes throughout Mexico to assess its usefulness for *Phaseolus* breeding due the
detection of six SCAR markers associated with common blight (*Xanthomonas axonopodis* pv.
phaseoli) resistance, as well as four for anthracnose (*Colletotrichum lindemuthianum*). The results
indicated significant morphological variability in the common bean germplasm. AFLP marker
analysis revealed high genetic diversity in those germplasms from north-central and central
Mexico. Germplasm from Morelos, Guanajuato, Querétaro, Durango, and Tamaulipas showed
the highest genetic diversity indexes. Cluster analysis was not consistent with classification
forms and their distribution based on geographical or agro-ecological origin. Germplasm from
Guanajuato and Tlaxcala showed the highest SCAR frequencies for both diseases. Genetic
diversity and SCAR detection for resistance to anthracnose and common blight was analysed
in two *P. coccineus* collections, one originating from the 'Huasteco Karst' located mainly in the
state of Puebla, Mexico. The other group of ayocote accessions comes from the state of
Veracruz. Analysis of germoplasm from Huasteco Karst revealed great genetic variability
among and within accessions as well as high genetic differentiation among germplasm.
Resistance to anthracnose was more frequent into the germplasm [31, 32]. Ayocote beans from

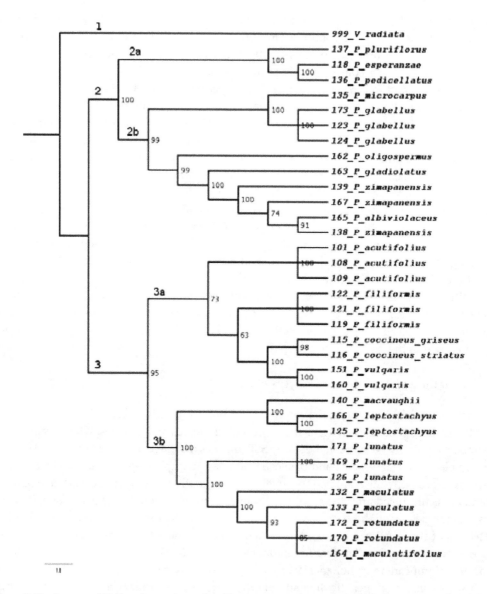

Figure 3. Dendrogram of 19 *Phaseolus* species analysed by gene sequencing and maximum parsimony method. Numbers indicate percentages of replication of node topology assessed by robustness method [76].

the state of Veracruz are highly variable, as those from Puebla; the domestication could have happened under reproductive isolation conditions and thus clearly differentiated lineages were produced based on the origin of each population. Since the ayocote beans analysed here include some accessions with high genetic variability, they are candidates for conservation and exploitation for *Phaseolus* breeding. Seed exchange in locations where ayocotes are cultivated through consumers and producers increase genetic diversity. Additionally, *P. coccineus* shows moderate open pollination (14.7%), which helps high genetic variation. The results indicated high genetic variability among and within accessions of ayocote beans from the states of Puebla and Veracruz; the data also suggest them to be an important source of allele useful for inclusion in breeding programmes of *P. coccineus* or *P. vulgaris* [33].

8. Concluding remarks

In order to understand genetic variation patterns and to reinforce the richness and genetic potential of *Phaseolus*, it is necessary to preserve, characterize and take advantage of Mexican germplasm. Knowledge of genetic diversity in the common bean could give us a better view for conservation as well as management and use of these plants' genetic resources. Despite significant effort and financial support by the Mexican government, this is not enough. Last year a new national centre for genetic resources was constructed and financed (Centro Nacional de Recursos Genéticos, CNRG, INIFAP) as well as initiatives to preserve not only beans but also other major Mexican crops such as maize, amaranth, cocoa, pumpkin, agave, avocado, etc. (SINAREFI-SAGARPA, Sistema Nacional de Recursos Fitogenéticos para la Alimentación y la Agricultura). We suggest that this financial support is maintained or improved, and research consistently supported through mid- and long-term projects.

Collection tours have demonstrated fast genetic erosion in most regions where wild types of *Phaseolus* were sought. Unfortunately, the social and economic situation in Mexico has provoked the migration of farmers and rural people, who have originally maintained and preserved genetic resources and their diversity in order to improve our lives. Thus, basic crops are leaving and are frequently changed by 'highly profitable' crops. The growth of urban populations has increased pollution and should have consequences through the alteration of natural reservoirs of natural populations inside 'Reservas de la Biósfera' or 'protected areas'. The Mexican government offers incipient economic support to those farmers that preserve genetic diversity *in situ* on their land. This strategy is appropriate for genetic resource conservation, but is currently not enough.

Genetic diversity observed in *Phaseolus* germplasm collections represent an important sample of total genetic variability contained in the genus. Molecular marker strategies and other recent and advanced techniques such as sequencing, genomics, proteomics and other '–omics' could help us to better understand genetic structure, genetic relations, genetic patterns of dispersion, and variation, promising germplasm to be used as parents or in breeding programmes based on their outstanding traits, etc. Biotechnology will not substitute traditional strategies of germplasm characterization or the taking advantage of *Phaseolus* germplasm. Biotechnology should be one allied or even more strengthen traditional breeding to accelerate and improve breeding methods and analysis strategies of genotypes.

Finally, we suggest that genetic diversity is a major challenge for botanists and taxonomists, biotechnologists and breeders, as well as to the governmental institutions of Mexico, towards maintaining natural populations both in situ and ex situ, to avoiding their loss, to increasing the strategies for their use, and to exploiting their benefits.

Acknowledgements

The authors are grateful for financial support provided by the Instituto Politécnico Nacional (project no. 1636), Fondo Mixto-Gobierno del Estado de Veracruz (project no. 94070) and Fondo CONACYT-Ciencia Básica (project no. 181756). Most of the information described here was derived from the MSc theses of V. M. Hernández-López, R. Ruiz-Salazar, V. H. Villarreal-Villagrán, and J. Martínez-Mondragón (CBG-IPN) and the PhD thesis of H. R. Gill-Langarica (CICATA-Unidad Altamira, IPN). SHD, JLChS, HRGL and NMP are S.N.I., EDI-IPN, and COFAA-IPN scholars. JMM was IPN-PIFI and CONACYT fellow.

Author details

Sanjuana Hernández-Delgado[1], José S. Muruaga-Martínez[2],
Ma. Luisa Patricia Vargas-Vázquez[3], Jairo Martínez-Mondragón[1], José L. Chávez-Servia[4],
Homar R. Gill-Langarica[1] and Netzahualcoyotl Mayek-Pérez[1*]

*Address all correspondence to: nmayek@ipn.mx

1 Centro de Biotecnología Genómica-Instituto Politécnico Nacional (CBG-IPN). Reynosa, Tamaulipas, Mexico

2 Instituto Nacional de Investigaciones Forestales, Agrícolas y Pecuarias (INIFAP). Texcoco, Mexico

3 INIFAP-Texcoco, Mexico

4 CIIDIR-Unidad Oaxaca, IPN, Oaxaca Mexico

References

[1] Mercado-Ruaro P, Delgado-Salinas A. Cytogenetics studies in *Phaseolus* L. (Fabaceae). Genetics and Molecular Biology 2000; 23: 985–987.

[2] Miranda S. Identificación de las especies cultivadas del género *Phaseolus*. Texcoco: Colegio de Postgraduados; 1990.

[3] Avendaño-Arrazate CH, Ramírez-Vallejo P, Castillo-González F, Chávez-Servia JL, Rincón-Enrique G. Diversidad isoenzimática en poblaciones nativas de frijol negro. Revista Fitotecnia Mexicana 2004; 27:31-40.

[4] Serrano J, Goñi I. Papel del frijol negro *Phaseolus vulgaris* en el estado nutricional de la población guatemalteca. Archivos Latinoamericanos de Nutrición 2004; 54: 36-44.

[5] Jaramillo S, Baena M. Material de apoyo a la capacitación en conservación *ex situ* de recursos fitogenéticos. Cali: Instituto Internacional de Recursos Fitogenéticos (IIRF); 2000.

[6] Rodrigo-Míguez AP. Caracterización morfoagrónomica y bioquímica de germoplasma de judía común (*Pasheolus vulgaris* L.) en España. PhD Thesis. Universidad de Santiago de Compostela. Pontevedra, Spain; 2000.

[7] Vargas MLP, Muruaga JS, Acosta JA, Navarrete R, Pérez P, Esquivel G, Irízar MBG, Hernández JM. Colección Núcleo de *Phaseolus vulgaris* L. del INIFAP. Catálogo de Accesiones de la Forma Domesticada. Chapingo, Mexico: Instituto Nacional de Investigaciones Forestales, Agrícolas y Pecuarias; 2006.

[8] Frankel OH, Brown AHD, Burgon JJ. Conservation of Plant Biodiversity. Sundridge UK: Cambridge University Press; 1995.

[9] Karmeswara-Rao N. Plant genetic resources: Advancing conservation and use through biotecnology. African Journal of Biotechnology 2004; 3: 138–145.

[10] Herrera-Estrella A. Deciphering the genome sequence of the Mesoamerican common bean. Proceedings Phaseomics –The Genome, June 21-26, 2012 Special Edition. Guanajuato, Mexico; 2012.

[11] Schmutz J, McClean PE, Wu GA, Cannon SB, Grimwood J, Jenkins J, Shu S, Song Q, Chavarro C, Torres-Torres M, Geffroy V, Moghaddam SM, Gao D, Abernathy B, Barry K, Blair M, Brick MA, Chovatia M, Goodstein DM, Gonzales M, Hellsten U, Hyten DL, Jia G, Kelly JD, Kudrna D, Lee R, Richard MMS, Miklas PN, Osorno JM, Rodrigues J, Thareau V, Urrea CA, Wang M, Yu Y, Zhang M, Wing RA, Cregan PB, Rokshar DS, Jackson SA. A reference genome for common bean and genome-wide analysis of dual domestications. Nature Genetics 2014; 46: 707–713.

[12] Jackson S. Genome sequence and analysis of Andean common bean G 19833. Proceedings Phaseomics –The Genome, June 21-26, 2012 Special Edition. Guanajuato, Mexico; 2012.

[13] Laboratorio Nacional de Genómica para la Biodiversidad (LANGEBIO). Institutional Projects. http://www.langebio.cinvestav.mx/?pag=170 (accessed 10 August 2014)

[14] Singh SP, Gutierrez JA, Molina A, Urrea C, Gepts P. Genetic diversity in cultivated common bean: II. Marker-based analysis of morphological and agronomic traits. Crop Science 1991; 31: 23–29.

[15] Servicio de Información Agroalimentaria y Pesquera (SIAP). Producción Agropecuaria y pesquera. Secretaría de Agricultura, Ganadería, Desarrollo Rural, pesca y Alimentación (SAGARPA). Mexico. http://www.siap.gob.mx (accessed 14 June 2014).

[16] Aparicio-Fernández X, Reynoso-Camacho R, Castaño-Tostado E, García-Gasca T, González- De Mejía E, Guzmán-Maldonado SH, Elizondo G, Gabra-Yousef G, Lila

MA, Loarca-Pina G. Antiradical capacity and induction of apoptosis on HeLa cells by a *Phaseolus vulgaris* extract. Plant Foods and Human Nutrition 2008; 63: 35–40.

[17] Carvalho MR, Sgarbieri VC. Relative importance of phytohemagglutinin (lectin) and trypsin-chymotrypsin inhibitor on bean (*Phaseolus vulgaris* L) protein absorption and utilization by the rat. Journal of Nutrition Science and Vitaminology (Tokyo) 1998; 44: 685–696.

[18] Gonzalez-De Mejia E, Valadez-Vega MC, Reynoso-Camacho R, Loarca-Pina G. Tannins, trypsin inhibitors and lectin cytotoxicity in tepary (*Phaseolus acutifolius*) and common (*Phaseolus vulgaris*) beans. Plant Foods and Human Nutrition 2005; 60: 137–145.

[19] López-Soto J, Ruiz JA, Sánchez JJ, Lépiz R. Adaptación climática de 25 especies de frijol silvestre (*Phaseolus* spp) en la República Mexicana. Revista Fitotecnia Mexicana 2005; 28: 221-230.

[20] Vargas-Vázquez MLP, Muruaga-Martínez JS, Pérez-Herrera P, Gill-Langarica HR, Esquivel-Esquivel G, Martínez-Damián MA, Rosales-Serna R, Mayek-Pérez N. Caracterización morfoagronómica de la colección núcleo de la forma cultivada de frijol común del INIFAP. Agrociencia 2008; 42: 787-797.

[21] Hernández-López VM, Vargas-Vázquez MLP, Muruaga-Martínez JS, Hernández-Delgado S, Mayek-Pérez N. Genetic diversity analysis of one collection of wild x cultivated bean accessions from Mexico. Bean Improvement Cooperative 2010; 53: 192–193.

[22] Hernández-López VM, Vargas-Vázquez MLP, Muruaga-Martínez JS, Hernández-Delgado S, Mayek-Pérez N. Detection of SCAR markers linked to resistance to anthracnose and common blight in wild x cultivated bean collection. Bean Improvement Cooperative 2010; 53: 214–215.

[23] Vargas-Vázquez MLP, Muruaga-Martínez JS, Martínez-Villarreal SE, Ruiz-Salazar R, Hernández-Delgado S, Mayek-Pérez N. Diversidad morfológica del frijol ayocote del Carso Huasteco de Mexico. Revista Mexicana de Biodiversidad 2011; 82: 767-775.

[24] Vargas-Vázquez MLP, Muruaga-Martínez JS, Lépiz-Ildefonso R, Pérez-Guerrero A. La colección INIFAP de frijol ayocote (*Phaseolus coccineus* L.) I. Distribución geográfica de sitios de colecta. Revista Mexicana de Ciencias Agrícolas 2012; 3: 1247-1259.

[25] Vargas-Vázquez MLP, Muruaga-Martínez JS, Pérez-Guerrero A. 2013. Temperatura y precipitación de los sitios de colecta de variedades nativas de frijol ayocote (*Phaseolus coccineus* L.). Revista Mexicana de Ciencias Agrícolas 2013; 4: 843-853.

[26] Vargas-Vázquez MLP, Muruaga-Martínez JS, Mayek-Pérez N, Pérez-Guerrero A, Ramírez-Sánchez SE. Caracterización de frijol ayocote (*Phaseolus coccineus* L.) del Eje Neovolcánico y la Sierra madre Oriental. Revista Mexicana de Ciencias Agrícolas 2014; 5: 191-200.

[27] Vargas MLP, Muruaga JS, Hernández JM, Díaz J. Los Recursos Genéticos Vegetales en el INIFAP: Estado Actual, Perspectivas y Desarrollo. Informe de resultados 2007. Mexico: INIFAP; 2007.

[28] Martínez-Castillo J, Zizumbo-Villarreal D, Perales-Rivera H, Colunga-Garcia, Marin P. Intraspecific diversity and morpho-phenological variation in *Phaseolus lunatus* L. from the Yucatán Peninsula, Mexico. Economic Botany 2004; 58: 354–380.

[29] Freytag GF, Debouck DG. Taxonomy, Distribution, and Ecology of the Genus *Phaseolus* (Leguminosae-Papilionideae) in North-America, Mexico and Central America. SI-DA, Botanical Miscellany 23. Fort Worth: Botanical Research Institute of Texas; 2002.

[30] Cárdenas F, Muruaga JS, Acosta JA. Catálogo: Banco de Germoplasma de *Phaseolus* spp. del Instituto Nacional de Investigaciones Forestales y Agropecuarias, Toluca: INIFAP; 1996.

[31] Ruiz-Salazar R, Hernández-Delgado S, Vargas-Vázquez MLP, Muruaga-Martínez JS, Mayek-Pérez N. Genetic diversity analysis of *Phaseolus coccineus* L. from Huasteco Karst of Mexico. Bean Improvement Cooperative 2010; 53: 182–183.

[32] Ruiz-Salazar R, Hernández-López VM, Hernández-Delgado S, Vargas-Vázquez MLP, Muruaga-Martínez JS, Mayek-Pérez N. Detection of SCAR markers linked to resistance to common blight and anthracnose in ayocote bean. Bean Improvement Cooperative 2010; 53: 212–213.

[33] Ruiz-Salazar R, Hernández-Delgado S, Vargas-Vázquez MLP, Muruaga-Martínez JS, Almaraz-Abarca N, Mayek-Pérez N. Genetic diversity of *Phaseolus coccineus* L. germplasm from Veracruz, Mexico. Bean Improvement Cooperative 2012; 55: 249–250.

[34] Acosta-Gallegos JA, Kelly JD, Gepts P. Prebreeding in common bean and use of genetic diversity from wild germplasm. Crop Science 2007; 47: S44–S59.

[35] Bitocchi E, Nanni L, Bellucci E, Rossi M, Giardini A, Spagnoletti Zeuli P, Logozzo G, Stougaard J, McClean P, Attene G, Papa R. Mesoamerican origin of the common bean (*Phaseolus vulgaris* L.) is revealed by sequence data. Proceeding of the National Academy of Sciences USA 2012; 109: E788–E796.

[36] Bitocchi E, Bellucci E, Giardini A, Rau D, Rodriguez M, Biagetti E, Santilocchi R, Zeuli PS, Gioia T, Logozzo G, Attene G, Nanni L, Papa R. Molecular analysis of the parallel domestication of the common bean (*Phaseolus vulgaris*) in Mesoamerica and the Andes. The New Phytologist 2013; 197: 300–313.

[37] Debouck DG, Toro O, Paredes OM, Johnson WC, Gepts P. Genetic diversity and ecological distribution of *Phaseolus vulgaris* (Fabaceae) in Northwestern South America. Economic Botany 1993; 47: 408–423.

[38] Cerón MS, Ligarreto GM, Moreno JD, Martínez O. Selección de variables cuantitativas y clasificación de 22 accesiones de frijol arbustivo (*Phaseolus vulgaris* L.). Revista Corpoica 2001; 3: 31-38.

[39] Rodiño AP, Santalla M, De Ron AM, Singh SP. A core collection of common bean from the Iberican peninsula. Euphytica 2003; 131: 165–175.

[40] Gill-Langarica HR, Muruaga-Martínez JS, Vargas-Vázquez MLP, Rosales-Serna R, Mayek-Pérez N. Genetic analysis of common bean core collection from Mexico. Genetics and Molecular Biology 2011; 34: 595–605.

[41] Rossi M, Bitocchi E, Bellucci E, Nanni L, Rau D, Attene G, Papa R. Linkage disequilibrium and population structure in wild and domesticated populations of Phaseolus vulgaris L. Evolutionary Applications 2009; 2: 504–522.

[42] Blair MW, Díaz LM, Gill-Langarica HR, Rosales-Serna R, Mayek-Pérez N, Acosta-Gallegos JA. Genetic relatedness of Mexican common bean cultivars revealed by microsatellite markers. Crop Science 2011; 51: 2655–2667.

[43] Blair MW, Soler A, Cortes AJ. Diversification and population structure in common beans (Phaseolus vulgaris L.). PLoS ONE 2013; 7: e49488.

[44] Blair MW, Díaz LM, Acosta-Gallegos JA. Race structure in the mexican collection of common bean landraces. Crop Science 2013; 53: 1517–1528.

[45] Desiderio F, Bitocchi E, Rau D, Rodriguez M, Attene G, Papa R, Nanni L. Chloroplast microsatellite diversity in Phaseolus vulgaris. Frontiers in Plant Science 2013; 3: 312.

[46] Okii D, Tukamuhabwa P, Kami J, Namayanja A, Paparu P, Ugen M, Gepts P. The genetic diversity and population structure of common bean (Phaseolus vulgaris L.) germplasm in Uganda. African Journal of Biotechnology 2014; 29: 2935–2949.

[47] Okii D, Tukamuhabwa P, Odong P, Namayanja A, Mukabaranga J, Paparu P, Gepts P. Morphological diversity of tropical common bean germplasm. African Crop Science Journal 2014; 22: 59–68

[48] Worthington M, Soleri D, Aragón-Cuevas F, Gepts P. Genetic composition and spatial distribution of farmer-managed Phaseolus bean plantings: an example from a village in Oaxaca, Mexico. Crop Science 2012; 52: 1721–1735.

[49] Santalla M, Menéndez MC, Monteagudo AB, De Ron AM. Genetic diversity of Argentinean common bean and its evolution during domestication. Euphytica 2004; 135: 75–87.

[50] Rodiño AP, Lema M, Pérez BM, Santalla M, De Ron AM. Assessment of runner bean (Phaseolus coccineus L.) germplasm for tolerance to low temperature during early seedling growth. Euphytica 2006; 155: 63–70.

[51] Sicard D, Nanni L, Porfiri O, Bulfon D, Papa R. Genetic diversity of Phaseolus vulgaris L. and Phaseolus coccineus L. landraces in central Italy. Plant Breeding 2005; 124: 464–472.

[52] Spataro G, Tiranti B, Arcaleni P, Bellucci E, Attene G, Papa R, Spagnoletti Zeuli P, Negri V. Genetic diversity and structure of a worldwide collection of *Phaseolus coccineus* L. Theoretical and Applied Genetics 2011; 122: 1281–1291.

[53] Rodriguez M, Rau D, Angioi SA, Bellucci E, Bitocchi E, Nanni L, Knüpffer H, Negri V, Papa R, Attene G. European *Phaseolus coccineus* L. landraces: Population structure and adaptation, as revealed by cpSSRs and phenotypic analyses. PLOS One 2013; 8: e57337.

[54] Martínez-Castillo D, Zizumbo-Villarreal D, Gepts P, Delgado-Valerio P, Colunga-GarciaMarin P. structure and genetic diversity of wild populations of lima bean (*Phaseolus lunatus* L.) from the Yucatán Peninsula, Mexico. Crop Science 2006; 46: 1071–1080.

[55] Martínez-Castillo J, Zizumbo-Villareal D, Gepts P, Colunga-García-Marín P. Gene flow and genetic structure in the wild–weedy–domesticated complex of *Phaseolus lunatus* L. in its Mesoamerican center of domestication and diversity. Crop Science 2007; 47: 58–66.

[56] Martínez-Castillo J, Camacho-Pérez L, Coello-Coello J, Andueza-Noh R. Wholesale replacement of lima bean (*Phaseolus lunatus* L.) landraces over the last 30 years in northeastern Campeche, Mexico. Genetic Resources and Crop Evolution 2012; 59: 191–204.

[57] Serrano-Serrano M, Andueza-Noh RH, Martínez-Castillo J, Debouck DG, Chacón Sánchez MI. Evolution and domestication of lima Bean (*Phaseolus lunatus* L.) in Mexico: evidence from Ribosomal DNA. Crop Science 2012; 52: 1698–1712.

[58] Martínez-Castillo J, Camacho-Pérez L, Villanueva-Viramontes S, Andueza-Noh RH, Chacón-Sánchez MI. Genetic structure within the Mesoamerican gene pool of wild *Phaseolus lunatus* (Fabaceae) from Mexico as revealed by microsatellite markers: implications for conservation and the domestication of the species. American Journal of Botany 2014; doi: 10.3732/ajb.1300412.

[59] Andueza-Noh RH, Serrano-Serrano M, Chacón Sánchez MI, Sánchez del Pino I, Camacho-Pérez L, Coello-Coello J, Mijangos Cortés J, Debouck DG, Martínez-Castillo J. Multiple domestications of the Mesoamerican gene pool of lima bean (*Phaseolus lunatus* L.): evidence from chloroplast DNA sequences. Genetic Resources and Crop Evolution 2013; 60: 1069–1086.

[60] Blair MW, Pantoja W, Carmenza-Muñoz L. First use of microsatellite markers in a large collection of cultivated and wild accessions of tepary bean (*Phaseolus acutifolius* A. Gray). Theoretical and Applied Genetics 2012; 6: 1137–1147.

[61] Molosiwa OO, Kgokong SB, Makwala B, Gwafila CM, Ramokapane MG. Genetic diversity in tepary bean (*Phaseolus acutifolius*) landraces grown in Botswana. Journal of Plan Breeding and Crop Science 2014; 6: 194–199.

[62] Schmit V, Debouck DG. Observations on the origin of *Phaseolus polyanthus* Green-man. Economic Botany 1991; 45: 345–364.

[63] Llaca V, Delgado A, Gepts P. Chloroplast DNA as an evolutionary marker in the *Phaseolus vulgaris* complex. Theoretical and Applied Genetics 1994; 88: 646–652.

[64] Hamann A, Zink D, Nagl W. Microsatellite fingerprinting in the genus *Phaseolus*. Genome 1995; 3: 507–515.

[65] Gaitán-Solís E, Duque MC, Edwards KJ, Tohme J. Microsatellite repeats in common bean (*Phaseolus vulgaris*): isolation, characterization and cross-species amplification in *Phaseolus* ssp. Crop Science 2002; 42: 2128–2136.

[66] Blair MW, Giraldo MC, Buendia HF, Tovar E, Duque MC, Beebe SE. Microsatellite marker diversity in common bean (*Phaseolus vulgaris* L.). Theoretical and Applied Genetics 2006; 113: 100–109.

[67] Benchimol L, de Campos T, Morais Carbonell SA, Colombo CA, Fernando Chioratto A, Fernandes Formighieri E, Lima Gouvea LR, Pereira de Souza A. Structure of genetic diversity among common bean (*Phaseolus vulgaris* L.) varieties of Mesoamerican and Andean origins using new developed microsatellite markers. Genetic Resources Crop Evolution 2007; 54: 1747–1762.

[68] Bellucci E, Bitocchi E, Ferrarini A, Benazzo A, Biagetti E, Klie S, Minio A, Rau D, Rodriguez M, Panziera A, Venturini L, Attene G, Albertini E, Jackson SA, Nanni L, Fernie AR, Nikoloski Z, Bertorelle G, Delledonne M, Papa R. Decreased nucleotide and expression diversity and modified coexpression patterns characterize domestication in the common bean. Plant Cell 2014; doi/10.1105/tpc.114.124040.

[69] Masi P, Logozzo G, Donini P, Spagnoletti P. Analysis of genetic structure in widely distributed common bean landraces with different plant growth habits using SSR and AFLP markers. Crop Science 2009; 49: 187–199.

[70] Chacón IR, Pickersgill B, Debouck DG, Salvador Arias JA. Phylogeographic analysis of the chloroplast DNA variation in wild common bean (*Phaseolus vulgaris* L.) in the Americas. Plant Systematics and Evolution 2007; 266: 175–195.

[71] Angioi, SA, Rau D, Attene G, Nanni L, Bellucci E, Logozzo G, Negri V, Spagnoletti Zeuli PL, Papa R. Beans in Europe: origin and structure of the European landraces of *Phaseolus vulgaris* L. Theoretical and Applied Genetics 2010; 121: 829–843.

[72] Ávila T, Blair MW, Reyes X, Bertin P. Genetic diversity of bean (*Phaseolus*) landraces and wild relatives from the primary centre of origin of the Southern Andes. Plant Genetic Resources 2012; 10: 83–92.

[73] Ramírez P, Castillo M, Castillo F, Miranda S. Diversidad morfológica de poblaciones nativas de frijol común y frijol ayocote del Oriente del estado de Mexico. Revista Fitotecnia Mexicana 2006; 29: 111-119.

[74] Hernández-López VM, Vargas-Vázquez MLP, Muruaga-Martínez JS, Hernández-Delgado S, Mayek-Pérez N. Origen, domesticación y diversificación del frijol común. Avances y perspectivas. Revista Fitotecnia Mexicana 2013; 36: 95-104.

[75] Larson G, Piperno DR, Allaby R, Purugganan M, Andersson L, Arroyo-Kalin M, Barton L, Climer Vigueira C, Denham T, Dobney K, Doust A, Gepts P, Gilbert T, Gremillion K, Lucas L, Lukens L, Marshall F, Olsen K, Pires C, Richerson P, Rubio de Casas R, Sanjur O, Thomas M, West-Eberhard MJ, Fuller DQ. Current perspectives and the future of domestication studies. Proceedings of the National Academy of Sciences USA 2014; 111: 6139–6146.

[76] Villarreal-Villagrán VH, Vargas-Vázquez MLP, Muruaga-Martínez JS, Hernández-Delgado S, Mayek-Pérez N. Phylogenetic analysis of *Phaseolus* spp. from Mexico. Bean Improvement Cooperative 2012; 55: 253–254.

[77] Delgado-Salinas A, Bibler R, Lavin M. Phylogeny of the genus *Phaseolus* (Leguminosae): a recent diversification in an ancient landscape. Systematic Botany 2006; 31: 779–791.

Evaluation of Genetic Diversity in Fish using Molecular Markers

Andreea Dudu, Sergiu Emil Georgescu and
Marieta Costache

1. Introduction

The variety of plants, animals and microorganisms on Earth, along with the tremendous diversity of genes in these species and the high variety of ecosystems on the globe are all constitutive parts of what is called "biodiversity". It is globally recognized that the conservation of nature and of its biological diversity should represent one of the priorities of the current human society. An important number of factors, sometimes interconnected, like climate change, pollution, anthropic activities destroying habitats for agriculture or logging, excessive exploitation of species with economic importance, hunting or poaching, represent serious treats for biodiversity.

The loss of biodiversity is a serious concern for multiple reasons, as a "healthy" biodiversity can provide various natural benefits including ecosystems services (contribution to climate stability, protection of water resources, nutrient storage and recycling, etc.), biological resources (food, medicinal and pharmaceutical resources, diversity in genes and species, breeding stocks and population reservoirs, etc.) and social benefits (research and education, recreation and tourism, etc.) [1].

Despite of well-known importance of biodiversity, human activity has been causing massive extinctions. In 2005, the report released by the Millennium Ecosystem Assessment pointed out a substantial and largely irreversible loss in the diversity of life on Earth, with some 10-50% of the well-studied higher taxonomic groups threatened with extinction, due to human actions. Over the past few hundred years, humans have increased species extinction rates by as much as 1000 times background rates that were typical over Earth's history [2]. Species are being lost at a rate that far exceeds the emergence of new species. The current extinction problem has been called the 'sixth extinction', as its magnitude compares with that of the other five mass extinctions revealed in geological records [3, 4].

The conservation of biodiversity at all levels from genes to ecosystems represents a global concern. Thus, the genetic diversity within species (between populations as well as among individuals within populations) as part of biodiversity is the result of the degree of variation at different levels (nucleotide, gene, chromosome, and genome). The presence of genetic variation plays an important role in species/populations survival and in their successful evolution in response to both short-term and long-term environmental changes [5]. To prevent the problems of genetic defects caused by inbreeding, species need a variety of genes to ensure successful survival. The decrease of this variety also translated as reduced genetic diversity is correlated with enhancing the chances of extinction.

The group of fish comprises over 32,000 species and an inestimable number of individuals inhabiting a large territory, from depths of 6000-7000 m on the bottom of oceans and seas to mountain waters found at an altitude of 2000 m. Thus, the fish group presents the most significant species variety among vertebrates [6]. The importance for humans derives mainly from economic significance, with aspects including consumption as food source, use in industry, aquaculture and fish farming. Overfishing and poaching along with habitat destruction (including water pollution, the building of dams, removal of water for use by humans, and the introduction of exotic species) represent the major threats to fish species and populations and currently, according to 2014 IUCN Red List, 2172 fish species are threatened with extinction [7].

Because of their life environment fish are more difficult to monitoring and study than terrestrial animal and to recover important information about fish populations is very challenging.

Sturgeons hold an important position in the category of the most threatened fish species in the world, mainly due to their particular scientific and commercial importance. Sturgeon species and the distantly related paddlefish as well as some extinct families are reunited in the Order Acipenseriformes and are generally regarded as "living fossils" and as the most primitive surviving bony fish. Sturgeons present some relict characters that differentiate them from other fish species and prove their ancient origin. The skeleton is primarily cartilaginous, with partial ossification only in cranium and maxilla, the posterior vertebrae continue far out into the dorsal lobe of the caudal fin (heterocercal condition), branchiostegal rays are absent or inconspicuous, undifferentiated vertebrae, ganoid scales only in caudal part of the body, five rows of bony scutes along the body. Sturgeons have a protrusible mouth, toothless in adults, a prominent snout with four barbells located about midway between the mouth and the snout tip [8].

Their extinct relatives (Infraclass Chondrostei) date to the Devonian (about 350 million years ago) and *Acipenser* fossils were found in the Western of North America dating from Upper Cretaceous. At present time within Acipenseriformes only two families exist, *Acipenseridae* and *Polyodontidae*, the first one counting 25 species inhabiting rivers, lakes, coastal waters and inner seas from the Northern hemisphere. The Acipenseriformes species are strictly Northern hemisphere fishes, some being resident to fresh water, while others are anadromous. Sturgeons are reunited in *Acipenseridae* family with four genera – *Acipenser, Huso, Scaphirhynchus* and *Pseudoscaphirhynchus*.

The area comprising the Black, Caspian and Aral Seas (Ponto-Caspian region) and the rivers that are tributaries to these seas, is the area with the greatest species diversity despite the fact the most of the populations in the region are facing extinction [9]. Currently in the Danube River three anadromous sturgeon species *Huso huso* (Beluga sturgeon), *Acipenser stellatus* (Stellate sturgeon) and *Acipenser gueldenstaedtii* (Russian sturgeon) and a potamodromous species *Acipenser ruthenus* (Sterlet sturgeon) are found [10]. The others two species that were present in the past in this area (*Acipenser sturio* – European sturgeon and *Acipenser nudiventris* – ship sturgeon) are considered to be extinct [11].

From ancient times, sturgeons had a great economic impact for the Danube region and they were an important part of the welfare of the local communities from this area. Due to human intervention, a decline of the sturgeon populations in this river has been observed starting with the 19th century, but much more accentuated in the 20th century. In Romania, in the first decade after the communist regime decline an overexploitation of sturgeon from Lower Danube has occurred mainly due to the absence of a legislation that protects these species [12, 13]. Thus, between 2002 and 2005 the sturgeon captures have severely diminished from 37 tons in 2002 to 11 tons in 2005. As a consequence of this evolution in 2006 the Romanian government adopted a law according to which the fishing of sturgeons for commercial purposes is prohibited for a period of 10 years. A solution to counter-balance the depletion of wild population was the one of sturgeon breeding in aquaculture. In the fish farms from Romania several species are preferred, especially Danube sturgeons as beluga sturgeon, stellate sturgeon and Russian sturgeon, considered to be ideal for obtaining caviar of superior quality and sterlet sturgeon that have the advantage of a very tasteful meat. Beside these species, the Siberian sturgeon (*Acipenser baerii*) and different inter-specific hybrids are also raised in aquaculture.

The causes for the dramatic decline of sturgeon populations all over the world are related mostly to the changes that are taking place in their habitats: pollution, dams and other constructions that are blocking the migration into the river for reproductive purposes, overfishing and poaching, mainly to provide caviar for the black market. Another factor that has a negative impact on the sturgeon populations is the practice of stocking the rivers with individuals from aquaculture, which becomes more and more popular with the years. Uncontrolled restocking presents as consequence the decrease of genetic diversity within and between populations, mainly due to the reduced number of adult individuals from the wild used as genitors for reproduction in aquaculture. Also, other practice with disastrous consequences for the populations is the one of restocking with inter-specific hybrids incorrectly labeled as individuals of native species. The inter-specific hybridization phenomenon is relatively frequent in sturgeons both in the wild and in aquaculture. Such hybrids were caught in the Danube River [14, 15] or are produced and raised in sturgeon farms. The species identifications based only on morphology might be misleading sometimes and molecular methods were proposed for species identification and caviar labeling [15-18].

Other fish species with a high ecological and economic importance are represented by salmonids. Romania is one of the European countries characterized by the presence of a significant number of wild salmonids populations and by an important potential for the development of intensive breeding of salmonids in aquaculture. The Salmoniformes order represents a heterogeneous group of fish reunited in one family *Salmonidae*, with three subfamilies (*Coregoninae* - whitefish and ciscos; *Thymallinae* – graylings; *Salmoninae* – trout, salmon and charr), nine genera and roughly 68 species [19]. The largest of salmoniform fishes are considered to be Chinook salmon (*Oncorhynchus tshawytscha*) and Danube huchen (*Hucho hucho*), that might reach at 1.5 m in length and a weight around 60 kg. Salmonids are characterized by body and fins that are streamlined and symmetrical, being covered with small and cycloid scales. All fins have soft rays. The representatives of this order possess several primitive anatomical features that are characteristics for an early stage in the evolution of modern bony fishes, like a small, fleshy adipose fin located between the dorsal fin and the powerful caudal fin [20].

They are native to the cooler climates of the Northern Hemisphere, but have been widely introduced around the world for angling and aquaculture.

In the Romanian fauna the following salmonid species are present: *Salmo trutta fario* (brown trout), *Salmo labrax* (Black Sea salmon), *Salvelinus fontinalis* (brook trout), *Hucho hucho* (huchen or Danube salmon), and *Thymallus thymallus* (grayling). The European whitefish (*Coregonus lavaretus maraenoides*) was introduced starting with 1957 in Lake Rosu, Bicaz and Tarcau rivers, the biological material originating from Poland and Russia. Unfortunately, there are no other studies regarding the adaptation of this species in the water systems where has been introduced. Apart the salmonids from the wild fauna previous mentioned the following species are bred for commercial purposes in aquaculture: brown trout, brook trout and rainbow trout (*Oncorhynchus mykiss*) and a series of hybrids of natural and fishery species.

The taxonomy of *Salmo* is still a matter of controversy [21] since more than 60 synonyms for varieties of brown trout and more than 20 for varieties of Atlantic salmon were described. The number of *Salmo* species recognized varies considerably not only because of highly phenotypic variation (body shape, colour, etc.), but also because the species inhabit and are adapted to very different habitats over large distribution areas. Due to taxonomic ambiguities, authors often refer to brown trout as *Salmo trutta* species complex [22, 23].

Salmo trutta comprises several distinct ecological and geographical morphs and with respect to this is still controversy as far as their classification as species or subspecies is concerned [24, 25]. Based on morphological and ecological variations, the existing populations of *Salmo trutta* from distinct areas are grouped into different taxa: i) Black Sea populations – *Salmo labrax*, ii) Caspian Sea populations – *Salmo caspius*, iii) Aral Sea populations – *Salmo oxianus* and iv) Mediterranean Sea populations – *Salmo macrostigma* [26].

The brown (common) trout (*Salmo trutta* morpha *fario* and *Salmo trutta* morpha *lacustris*) and the sea trout (*Salmo trutta* morpha *trutta*) are fish of the same species, considered by some taxonomists different subspecies in order to distinguish the anadromous *Salmo trutta trutta*, living in the sea and migrating in freshwater only to spawn, from *Salmo trutta fario*, residing in freshwater and the lake dwelling form *Salmo trutta lacustris*. Instead, other authors consider that these do not necessarily represent monophyletic groups [27].

In Romania, *Salmo trutta fario* (Linnaeus, 1758) is widely spread in a large number of water streams from the mountain area, along the Carpathian Arch, whereas the Black Sea trout, *Salmo labrax* (Pallas, 1814) is endemic to the Black Sea area and migrates for reproduction in the Danube River and its tributaries. Nowadays only few individuals are captured annually along the Black Sea coast and sporadically in the Lower Danube River. Due to the present situation of the species several measures were adopted to protect it. Thus, the fishing is completely prohibited and the Black Sea trout is included on the Red List of Danube Delta Biosphere Reserve [8].

Salvelinus fontinalis (Mitchill, 1815) is predominantly raised in fish farms for food consumption, but a low number of wild populations are still present in the Romanian mountain waters.

Thymallus thymallus (Linnaeus, 1758) is the only native salmonid species for which no imports of biological material and restocking programs were completed in Romania [26].

Until two or three decades ago, excellent habitat conditions for Danube huchen (*Hucho hucho*, Linnaeus, 1758) still existed in many rivers in Romania. The historical range of this species included the majority of the Carpathian river systems [28], but hydropower development, river pollution, and overfishing and poaching led to drastic declines in the area inhabited by huchen [29], and the species is now extinct in the Mures, Timis, Cerna, Olt, Arges, and Ialomita river systems. Data on the current area of huchen occurrence in Romania are fragmentary, and the species only occurs in a few rivers, including, among others, the Tisa River and its tributaries (Viseu, Ruscova, Crasna, Bistra, Vaser, Somes, and Cris) and in the Siret River.

In relation to the main rivers from Romania, there are 10 hydrographic basins in which significant salmonid populations are found. Nowadays is a well-known fact that the most prevalent salmonid species is the brown trout, while the most threatened are the Danube huchen and the Black Sea salmon.

Both sturgeons and salmonids are of crucial importance for Romanian fauna, having also a significant socio-economic value. In this context, well documented studies regarding the biology, taxonomic classification and ecology of the species were performed [10, 28, 30-32]. The development of technologies based on DNA markers has had a tremendous impact on animal genetics in generally, and changed the way in which studies were conducted inclusive in population genetics, phylogeny, phylogeography and conservation. Thus, in a progressive way, studies aiming at analyzing molecular aspects in sturgeons and salmonids populations from Romania founded their place in the research field, although the number of such studies is still relatively low.

2. Molecular markers for population genetics and conservation

The definition of "marker" in wide sense is "something that serves to identify, predict, or characterize" [33]. In biology, the markers refer to any stable variation, which is heritable and can be measured or detected by an appropriate method. Such variations are produced at different levels for example, morphological, gene, chromosomal, biochemical or genomic. The markers that represent variations that appear at the DNA level are so-called *molecular mark-*

ers. The molecular markers occupy specific places in the genome and possess the role to "mark" the position of a specific gene or the inheritance of a particular character.

All organisms are subject to mutations that appear when genetic material fails to copy accurately or as a consequence of interactions with the environment, leading to genetic variation or polymorphism. Individuals of a species are distributed into more or less separate groups called populations, distributed over the species range. The genetic variation of a species is distributed both within and between populations. The diversity at the gene level is often referred to as intra-specific variability as it represents biological diversity within a single species. The researchers are very interested in assessing genetic variation within and between populations and detecting similarities as well as differences between individuals/populations in order to establish optimum conservation strategies.

For this variation to be useful in different genetic studies it is necessary to be inherited and to be perceptible for the researchers, whether is recognizable as phenotypic variation or as a genetic mutation detectable by different molecular techniques. At the DNA level the genetic variation is represented by point mutations called also single nucleotide polymorphisms (SNPs), indels (insertions or deletions of nucleotide sequences), inversion of a segment of DNA within a locus, and rearrangement of DNA segments around a locus of interest. As the mutation rate is very low the evolution through mutation is extremely slow. The process of mutation is the only way in which genetic variation is created and in the lack of mutations there would be no biological diversity.

DNA marker technology can be applied to reveal these mutations. Large deletions and insertions determine shifts in the size of DNA fragments resulted consequently digestion by restriction enzymes, and are among the easiest type of mutations to detect by electrophoresis on agarose gel; smaller indels require DNA sequencing or more elaborate electrophoretic techniques, while the SNPs can be easily detected by DNA sequencing [34].

By using molecular markers is possible to observe and exploit genetic variation across the entire genome. Thus, the application of molecular markers in fish allowed recording rapid progress regarding the study of genetic variability and inbreeding, parentage determination, species identification, genetic linkage map construction for aquaculture species, identifying **Quantitative Trait Loci (QTL)** related to specific traits for marker assisted selection. One of the classifications of the molecular markers refers to **markers of type I** as markers associated with genes having a known function and **markers of type II** as markers associated with unknown genomic segments [35]. The type I of DNA markers were not initially consider to be appropriate for genetic studies in fishes, but over time it became clear that these markers are very important both for the study of wild populations and aquaculture. Thus, this of markers has become very important for studying the phenomenon of linkage and for QTL mapping, being of great use in comparative genomic studies and for identification of candidate genes for quantitative traits in different fish species raised in aquaculture. An important number of studies were focused on the elucidation of the molecular basis of economically important traits in different species of salmonids. For example in the case of Atlantic salmon the mapping the QTLs was performed for the loci correlated with viral disease resistance [36], flesh color and growth traits [37], salinity tolerance [38], late sexual maturation [39], etc.

The type II of molecular markers (**RAPD, AFLP, microsatellites**) is considered to be non-coding. Such markers are used in population genetic studies aiming to characterize the genetic diversity. The type II of DNA markers proved useful in identifying species, populations and subpopulation, but also in identifying interspecific hybrids.

A second classification of markers is the one regarding their position in the cell. Thus we can distinguish between **nuclear and mitochondrial DNA (mtDNA) markers**, depending on their localization in the nuclear or mitochondrion genome.

The molecular markers might be highlighted by a variety of techniques that differ by difficulty, repeatability, cost and nature of polymorphism that is detected. The **DNA markers** detected by Polymerase Chain Reaction (PCR) are the most frequent used in assessing the genetic diversity and have a particular role in conservation. By PCR is possible to amplify DNA sequences up to several million times, so these markers present the advantage of nonlethal sampling. Small amount of tissue (fin clips or scales) are sufficient for analysis and is not necessary that the individuals to be sacrificed for sampling. This can be an important feature when evaluating genetic change in protected or declining populations and for providing access to DNA of ancient or archived tissue samples. It can provide information about genetic diversity over extensive temporal and spatial scales, especially for populations that no longer exist. Taken together, these characteristics suggest capability to monitoring populations that are small, exploited or declining.

Depending on the primers used for PCR amplification these markers can be divided in two groups: **(i) PCR markers for target sequences** – in this case the fragment of interest is amplified with two specific primers and **(ii) PCR markers for arbitrary sequences** – one primer with a arbitrary nucleotide sequence is used; the primer binds to randomly in the genome resulting unknown DNA fragments. The primer used in this type of techniques is usually short (of 8-10 nucleotides) and so the probability for it to bind at multiple sites in genome increase.

In the first category are found molecular markers like PCR-RFLP (PCR Restriction Fragment Length Polymorphism), PASA (PCR Amplification of Specific Alleles), SNP (Single Nucleotide Polymorphism), repetitive DNA sequences (minisatellites, microsatellites, etc.) [34]. The second category includes markers like RAPD (Randomly Amplified Polymorphic DNA), AP-PCR (Arbitrary Primed-PCR) and AFLP (Amplified Fragment Length Polymorphism), but these are less preferred comparing with PCR markers for target sequences due to the difficulty of analysis and lack of results accuracy and reproducibility.

PCR-RFLP

The analysis of these markers involves the amplification by PCR of a specific DNA region comprising one or more polymorphic sites for restriction enzymes (RE). With the increasing number of so-called "universal" primers available, can be targeted DNA regions that are relatively conserved among species. In addition, PCR products can be digested with restriction enzymes and visualized by ethidium bromide staining due to the increased amount of the resulting DNA amplification reaction. The ability of RFLP markers to highlight the genetic variation is relatively low compared with the one of other markers. Substitutions, insertions, deletions and rearrangements of the regions containing restriction sites are probably quite widespread in the genomes of many species, but the probability that they exist at the locus that we want to study is quite low. Because the difference in size between the restriction fragments is usually large these can be easily separated by agarose electrophoresis. The disadvantage of PCR-RFLP is that presents a relatively low level of polymorphism and requires knowing the sequence of the fragment amplified by PCR, fact that makes difficult to establish new markers.

In genetic studies regarding sturgeons the PCR-RFLP markers were mainly used in two directions: to assess the genetic diversity of sturgeon populations [40, 41] and for species identification and implicit for caviar traceability [42, 43]. Wolf et al. [42] has identified species specific restriction profiles in a fragment from the mitochondrial gene *cytochrome b* (*cyt b*) in 10 species from *Acipenser* and *Huso* genera. A similar study was performed by Ludwig et al [44] for 22 sturgeon species, an accurate identification being possible by analyzing the restriction profiles resulted consequently the action of five REs on a fragment of 1121 bp from *cyt b* gene. Panagiotopoulou et al. [43] proposed a molecular method based on PCR-RFLP to distinguish between Atlantic (*Acipenser oxyrinchus*) and European (*A. sturio*) sturgeon. The discrimination between the two species is difficult to be done exclusively by morphological traits, especially in their early life stages, while the application of two REs allowed the clear and unambiguous discrimination of 132 specimens of Atlantic and European sturgeon.

The identification of acipenserid species by PCR-RFLP presents the difficulty of correct diagnostic for species that are closely related from genetic and evolutionary point of view, like is the case of *A. gueldenstaedtii* / *A. persicus* or *Scaphirhynchus* genus species. Moreover, the analysis of length polymorphisms for mitochondrial DNA fragments put in evidence only the genetic variability originating from the maternal genitor, as the mtDNA is almost exclusively maternally inherited. This fact might lead to misinterpretations in case of inter-specific hybrids diagnostic and/ or an ancient introgression, both phenomena occurring in sturgeons [15, 45].

PCR-RFLP markers were used also in salmonid species for the genetic diversity and phylogeography analyses in different populations of brown trout [46, 47], charrs [48], grayling [49], etc.

Microsatellites also referred to as "simple sequence repeat" (SSR) [50] represent short repetitive sequences of 2-9 bp, wide spread in the genome and with a significant level of polymorphism. These markers are numerous in vertebrates, in fish appearing in every 10kb [51]. The majority of microsatellite loci are relatively small sizes, being amplified easily by PCR. Up to 70% of loci present dinucleotide repeats, the (AC) motif being the most common in the vertebrate genome [52]. The main features of microsatellites are co-dominant inheritance, high degree of polymorphism, hypervariability, higher mutation rate than standard. Through microsatellites analysis is possible to infer the genetic profile of an individual (genetic fingerprint) and to establish the relationships between individuals. Some microsatellites have a high number of alleles for a locus within population and are very suitable to identify the genitors and their progeny in hybrid populations. The higher level of allelic variation at microsatellite markers make them useful for addressing questions related to genetic structure, particularly where genetic differentiation may be limited. On the contrary other microsatellites present a low number of alleles or even a fixed allele and are more appropriate for phylogeny or species identification. The utility of microsatellite markers can be determined depending on the polymorphic information content (PIC), defined as the capacity of the marker to detect a polymorphism in the population [53]. PIC is direct correlated with the number of detected alleles and their frequency, a higher number of alleles determining a higher value for PIC. By comparing the values for PIC in different markers is possible to obtain valuable information about their power and efficiency in population genetic and conservation studies.

The primers designed for microsatellites amplification in a species give also cross-amplification for similar loci in related species, this being an important benefit in analyzing populations that are small or at the brink of extinction [54]. Microsatellite markers are widely used for population genetic and conservation studies in fishes.

The first studies based on microsatellite loci analysis were initiated at the North-American species, where it aimed at isolation and description of disomic loci [55, 56, 57].

The identification and characterization of new loci is complicated by the polyploidy of the sturgeon species. Lots of potential useful microsatellites were eliminated from analysis since these were polysomic and thus, they complicated the interpretation of the inheritance mode and of genetic variation within and between populations. Once a disomic set of microsatellites is established, this fact permits to analyze the genetic diversity and structure of wild populations and aquaculture stocks. Such disomic loci were isolated and characterized also for *A. naccarii* [58, 59] and *A. persicus* [60]. In the case of Ponto-Caspian sturgeons, a few studies based on microsatellites analysis in *A. stellatus* from the Caspian Sea [61] and, respectively, in *A. gueldenstaedtii* from the North-West of the Black Sea, North of the Caspian Sea and Azov Sea [62] were performed.

Beside their high applicability in for inferring the genetic diversity and structure of the wild populations, the microsatellites are appropriate for aquaculture stocks evaluation, selection of breeders and proper conservation. Different studies were conducted in several sturgeon species (*A. transmontanus*, *A. fulvescens*, *A. naccarii*, *A. sturio*, etc.) for *ex-situ* conservation purposes. The obtained data were used in the management program to adopt appropriate conservation methods.

In salmonids, the microsatellites were successfully applied in phylogeography studies [22], in determination of genetic variation in wild and farmed fish populations [63], in inferring the genetic diversity within population, fine-scale genetic differentiation and relationship of populations [64], in assessment of stocking impact on wild populations [65].

The mitochondrial genome possesses certain characteristics (compact organization, maternal inheritance, hundreds to thousands of copies per cell, rapidly evolving, reduced recombination rate, and higher mutation rates compared to those of nuclear genes) that make it useful in population genetics and phylogeny studies. mtDNA analysis in sturgeons is a suitable method for the characterization of species and populations, providing useful information for the management of conservation activities. For intraspecific studies, the most commonly used marker is the mitochondrial control region (D-loop) due to the relatively high degree of nucleotide variation. Studies to discriminate species/populations/aquaculture strains based on the analysis of D-loop region polymorphisms were performed for several species of sturgeon, such as *A. gueldenstaedtii* [66], *A. sinensis* [67], *A. stellatus* and *A. baerii* [68].

The gene coding for *cyt b* is a useful marker for identifying different species of sturgeon based on specific polymorphisms found at this level. A 648 bp fragment of the gene coding for *cytochrome oxidase subunit I (CO I)* is considered a real "barcode" for vertebrates and is useful for species identification by *DNA barcoding* technique.

Genes with highly conserved sequence and slow evolution like mitochondrial ribosomal genes (16SrRNA and 12SrRNA genes) are preferred when is about inferring the phylogenetic relationships of fishes at different taxonomic levels. 12SrRNA gene is considered a promising tool for tracing the history of more recent evolutionary events and it has been widely used to study the phylogenetic relationships among different levels of taxa such as families, genera and species [26].

In salmonid species mtDNA has proven to be useful for phylogeographical studies [69] and for analysis of spatial and temporal population structure [70]. Based on the analysis of the mitochondrial D-loop marker Bernatchez [22] showed that there are five different main lineages of *Salmo trutta* in Europe: Atlantic, Danubian, Marmoratus, Mediterranean and Adriatic. Analytical techniques for mtDNA include indirect methods such as the analysis of RFLP markers or the direct analysis of mtDNA sequences.

3. Factors influencing the genetic diversity of populations

The genes are transferred from one generation to the next, and every individual has two copies of each gene, one which is inherited from the maternal genitor, the other from the paternal genitor. The DNA sequence of a specific locus, either a gene or a non-coding marker, may present some differences, resulting different variants of the same locus. Such variants of a specific locus are called alleles and their existence implies the genetic variation existence.

i.   **Mutation** is in wide sense the process of random change of the DNA sequence and represents a process by which new alleles are created. As the mutation rates at nuclear and mitochondrial genome are very slow, the evolution through mutation is so slow that it is generally impossible to detect it from one generation to another. However, mutation is important as a source of genetic variation. The process of mutation is the only way in which genetic variability is created, and without mutations there would be no biological diversity.

ii.   **Gene flow.** A population acquires new alleles mostly through the immigration of individuals from surrounding populations (gene flow) and through mutations. Apart from these two processes the number of alleles in a population is determined by the size of the population. Gene flow is a change in allele frequency that occurs due to migration of individuals among populations. It is possible that individuals that are moving into a new population to bring new alleles which are not present in that population or that they are in frequencies that differ from the allele frequencies of that population. So the gene flow increases the genetic variation within a population, but tends to make populations genetically similar to each other. If migration between populations occur in large numbers of individuals and the level of gene flow is significantly high, the populations will have the same alleles in the same frequencies and it will be one single population.

iii.   **Genetic drift** represents a random modification in allele frequency of population that occurs if a population size is not infinite. In populations that are of limited size, allele frequencies will change randomly from one generation to the next. In the short term, over a few generations, a result of genetic drift would be the increasing or decreasing of allele frequencies in a random, unpredictable way. In the longer term, the main result of genetic drift is loss of genetic variation as by chance some of the alleles that exist in the parent generation may not be passed on to their offspring. The effects of genetic drift are strongest in small populations because the more impressive is the fluctuation of allele frequencies, and the sooner the loss of genetic variation. Genetic drift also results in different populations becoming genetically different from each other because different alleles will become more frequent or fixed in different populations.

iv.   **Natural selection** is the gradual process by which biological traits become either more or less common in a population as a function of the effect of inherited traits on the differential reproductive success of organisms interacting with their environment. Natural selection appears because different genotypes have different fitness. Individuals with higher fitness survive and reproduce more than other individuals, so these genotypes become increasingly more and more frequent in populations. In different populations, parents of different genotypes pass their genes unequally to the next generation, leading to the genetic differences among isolated populations. So, genetic drift tends to make different populations genetically distinct from each other by chance, whereas natural selection tends to form genetically different populations due to environmental constraints. Consequently, the traits that have high fitness in one population, and evolve through natural selection, will be different from the traits that have high fitness and evolve through natural selection in another population. In generally, the natural selection, genetic drift and gene flow have an effect on genetic variation within populations and between populations. While the genetic drift and selection tend to reduce the variation within populations and increase the differences between populations, the gene flow increases the variation within populations, but makes populations similar.

v.   **Inbreeding.** The birth of offspring resulted from reproduction between close relatives that occurs mainly in small and isolated populations. The consequence of inbreeding is the reduced viability and reproduction, as well as increased occurrences of diseases and defects, so called inbreeding depression [71].

4. Molecular studies for species identifications and genetic diversity assessing in sturgeons from Romania

The markers analyzed in sturgeon species and populations from Romania were nuclear markers (microsatellites) and mtDNA. The methodologies applied in the studies included microsatellites genotyping, PCR-RFLP and sequencing of mitochondrial markers. Thus, the DNA was isolated from biological samples consisting in small pieces of fin sampled without harming the animals (aspect very important in vulnerable populations) by a classic protocol with phenol-chloroform-isoamylic alcohol.

For mtDNA analysis the primers were designed based on DNA sequences for the interest markers retrieved from GenBank data base. The primers for microsatellites amplification were described in literature in different species and the cross-amplification for the similar species from Romania was tested.

The molecular markers were amplified by PCR or multiplex PCR in specific conditions established consequently several steps of reaction optimization. The microsatellites were analyzed by capillary electrophoresis in ABI Prism 310 Genetic Analyzer (Applied Biosystems) and the mitochondrial markers were sequenced by Sanger method, *dye terminator* variant, in ABI Prism 3130 Genetic Analyzer (Applied Biosystems). The raw data were processed and edited with dedicated computer programs. The interpretation of genotypic data was done by using specialized computer software for population genetics and phylogeny.

In the context of severe decline of sturgeon population from the Lower Danube several studies based on molecular marker analysis were performed in the recent years in Romania. In consequence, the studies were directed to assessing the genetic diversity in Lower Danube sturgeon populations based on microsatellites and to evaluate genetic variability in aquaculture strains in order to sustain the efforts of conservation.

The analyses regarding the genetic diversity were preceded by the correct species identification based on molecular markers for each individual. As mentioned before, the accurate detection of sturgeon species encountered in the Danube River can be very difficult, due to the plasticity of various external morphological features. The number of hybrid individuals occurring in natural conditions is unknown. However, the number of hybrids in natural waters could increase due to escapes from commercial farms where exotic species or genotypes and inter-specific hybrids are regularly used for production purposes. So, decisions to initiate restocking programs for sturgeon conservation need to take into account, besides the socio-economical aspects and the assessment of genetic diversity, the correct diagnostic of fishes included in this type of programs.

For sturgeon species from the Lower Danube diagnostic a molecular methods was set-up by Dudu et al [15] based on the analysis of nuclear markers. In this study 84 individuals belonging to all four species from the Lower Danube (A. *stellatus, A. gueldenstaedtii, A. ruthenus* and *H. huso*) were analyzed. Initially 25 microsatellites were tested from which eight loci (LS19, LS34, LS39, LS54, Aox27, AoxD234, AnacE4, and AnacC11) that have demonstrated a good amplification, results repeatability and interspecific polymorphism in all of the four investigated species were selected. The genotypic data processing involves three successive statistical analyses including Factorial Correspondence Analysis (FCA), STRUCTURE assignation and NewHybrids status determination.

In a first step, the genotypes data were run in a FCA test, using GENETIX software only for the individuals considered being pure species based on morphology analysis. The FCA highlighted the differences between the four analyzed species. Four main clusters, each corresponding to one of the sturgeon species analyzed, were identified (Figure 1).

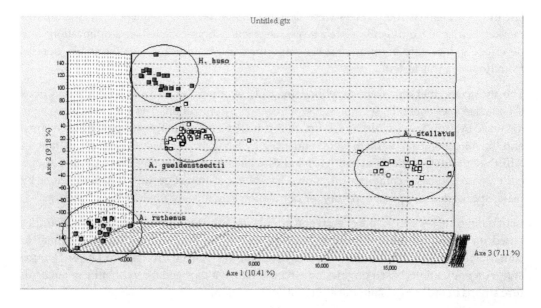

Figure 1. Factorial Correspondence Analysis (FCA) based on 8 microsatellite loci in the four pure species of Danube sturgeons [15].

In the second step, the genotypes data of putative hybrids, diagnosed such as based on morphometric indices were included in FCA analysis as supplementary individuals. A fifth category was highlighted including the putative hybrids, but also individuals that were considered as pure species, based on their morphology. The analysis had showed that only three of the four species hybridize each other, the putative hybrids displaying an intermediate position between *A. stellatus*, *H. huso* and *A. gueldenstaedtii*. Consequently, we deduced that *A. ruthenus* individuals should be eliminated from further tests (Figure 2).

In consequence in the third step, FCA was performed only on three pure species and hybrid individuals included as supplementary elements in analysis. The analysis grouped the pure species and the hybrids in distinct clusters. The hybrids and some of the individuals labeled as pure were occupying an intermediate position between the pure species groups, some of these being placed approximately in the middle of the triangle delimited by the three pure species, while others appear to be closer to *H. huso* and *A. gueldenstaedtii* (Figure 3).

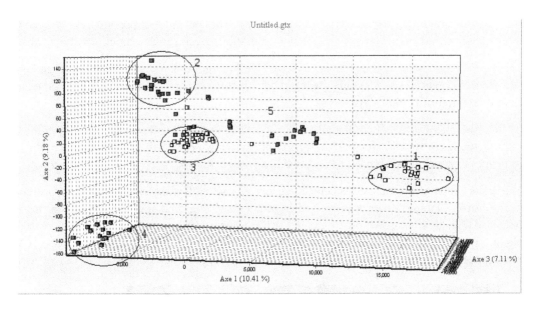

Figure 2. Factorial Correspondence Analysis (FCA) based on 8 microsatellite loci in the four pure species and hybrids of Danube sturgeons. (1) *A. stellatus*; (2) *H. huso*; (3) *A. gueldenstaedtii*; (4) *A. ruthenus*; (5) Hybrids and some pure individuals to be tested with other methods [15].

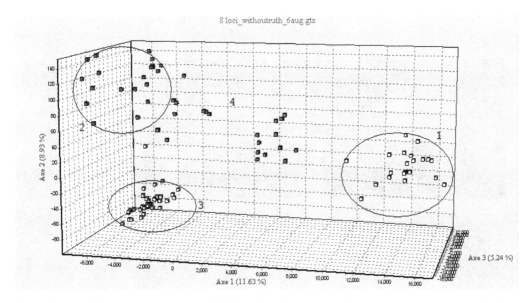

Figure 3. Factorial Correspondence Analysis (FCA) based on eight microsatellite loci in three pure species and hybrids of Danube sturgeons: (1) *A. stellatus*; (2) *H. huso*; (3) *A. gueldenstaedtii*; (4) Hybrids and some pure individuals [15].

The second statistical analysis of the genotype data was the assignment test with STRUCTURE, which confirmed the presence of five specific clusters: (1) *A. stellatus* pure species; (2) *H. huso* pure species; (3) *A. gueldenstaedtii* pure species; (4) *A. ruthenus* pure species; (5) hybrids (Figure 4). The individuals classified as pure based on morphology (with three exceptions) were strongly assigned in their corresponding species. The perfect correspondence between morphological and molecular/assignation determinations gives a great reliability to the results.

The hybrids confirmed by the FCA and STRUCTURE assignment test were analyzed together with their two genitor species using the NewHybrids software in order to distinguish between F1 and later hybridization steps.

By the three successive statistical analyses it is possible the diagnostic of surgeon individuals as pure species or hybrids, the method proposed showing a high efficiency in discriminating pure species specimens from F1, F2 and two kinds of backcross.

The molecular analysis for species detection is necessary for all sturgeon individuals captured in the Lower Danube River and implied in restocking programs and from aquaculture. A database with genotypic data can be created as the accuracy of the proposed method increases along with the number of reference individuals analyzed. Also, such database might be of a real use for designing efficient management plans for sturgeon populations.

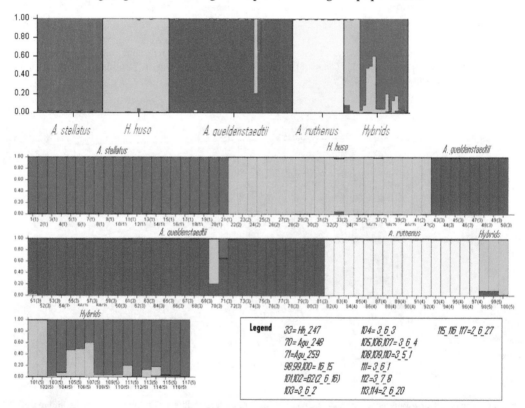

Figure 4. Assignation of the 84 sturgeons by STRUCTURE analysis based on eight microsatellite loci in four sturgeon species from Danube. Histograms represent the estimated membership coefficients (Q). Composite bars are expected hybrids [15].

When a hybrid is detected the origin of maternal species can be identified by analyzing mitochondrial markers. For example the analysis by sequencing or PCR-RFLP of *cytb* gene permits the identification of the maternal genitor involved in hybrid formation as is well known that the mitochondrion genome is exclusively maternally inherited. The species identification based on the analysis of mtDNA by different methods should be handled with precaution when we deal with species that can easily hybridize, like sturgeons. The mito-chondrial marker analysis is suitable for pure species, but totally inefficient for hybrids.

The analysis of genetic diversity in the populations from the Lower Danube by microsatellites was hindered by the complexity of the genome in this group of fish.

All Acipenseriformes are divided into three separate groups depending on the number of chromosomes: (1) species with karyotypes comprising about 120 chromosomes; (2) species with 240 to 270 chromosomes; they are conventionally referred to as 250-chromosomes species; (3) species with around 370 chromosomes [72]. Two scales of Acipenseriformes ploidy have been proposed: (1) the "evolutionary scale": diploid species (extinct), tetraploid species (120-chromosomes), octoploid (250 chromosomes), and 12-ploid (370-chromosomes) species [73]; and (2) the "contemporary scale": diploid (120-chromosomes), tetraploid (250-chromosomes), and hexaploid (370-chromosomes) species [74].

Three of the four sturgeon species from the Lower Danube – *A. stellatus, A. ruthenus* and *H. huso* are considered to be functional diploid as the process of functional reduction of the genome is consider being almost completed in these species, while *A. gueldenstaedtii* is functional tetraploid, with an octaploid ancestor from which some loci are maintained. The analysis of a set of 12 microsatellites loci (AciG93, AciG198, AnacC11, AnacE4, Aox27, AoxD234, As002, LS19, LS34, LS39, LS54, Spl106) in 51 individuals of *A. gueldenstaedtii* from Lower Danube revealed that only two loci were disomic, while the others were polysomic with a number of 3-8 alleles per locus in an individual (Table 1).

Locus	Allele size (bp)	Maximum number of alleles per individual
AciG93	383-399	4
AciG198	184 - 188	2
AnacC11	144-204	4
AnacE4	320-356	3
Aox27	110-142	4
AoxD234	188-268	4
As002	92-140	8
LS19	115-154	4
LS34	121-151	4
LS39	85-157	4
LS54	117-237	4
Spl106	234-246	2

Table 1. Characteristics of 12 microsatellite loci in *A. gueldenstaedtii* from Lower Danube.

The analysis of genetic diversity by using microsatellite with polysomic pattern is facing the problem of correct determination of the genotype. For example, for a locus with tetrasomic pattern for which only three peaks are detected by capillary electrophoresis is hard to determine which allele has two copies. The method of *gene dosage* proposed by Jenneckens et al [75] by which the genotype can be determined by calculating the report of peak areas has proven to be inefficient in our case.

For the *A. stellatus, A. ruthenus* and *H. huso* population from the Lower Danube genetic diversity studies were performed based on microsatellite analysis. From an extended set of microsatellites isolated originally in the North–American sturgeon species *A. fulvescens* and *A. oxyrhinchus* and in Adriatic sturgeon – *A. naccarii*, seven loci (LS19, LS34, LS39, LS54, AnacE4, AnacC11, and AoxD234) that showed good results of amplification and a disomic pattern in all the three species of sturgeon from Lower Danube, were selected for the assessment of genetic diversity.

The analysis was performed in a total number of 158 sturgeon individuals (62 of *A. stellatus*, 54 of *H. huso* and 42 of *A. ruthenus*) that were captured in the river between 2001-2008 as part of a national scientific research study for restocking and monitoring.

The estimation of the genetic diversity was realized by inferring several statistic indices with Genetix software. Thus the average values of expected heterozygosity (H_E), observed heterozygosity (H_O) and the mean number of alleles/locus (MNA) were calculated.

The heterozygosity is the percentage of heterozygous loci in a population. The average values of H_O are similar for the three sturgeon population from the Lower Danube (0.431 – *A. stellatus*; 0.476 – *H. huso*; 0.4017 – *A. ruthenus*), while the average values of H_E are situated between 0.6409 (*H. huso*) şi 0.5634 (*A. ruthenus*). The highest values of H_O (0.4760) and MNA per locus (6.2875) were obtained for *H. huso*, which appears to have the highest variability among the analyzed populations.

Locus	Population								
	A. stellatus			*H. huso*			*A. ruthenus*		
	H_E	H_O	MNA	H_E	H_O	MNA	H_E	H_O	MNA
LS19	0.7053	0.5484		0.8033	0.6667		0.4963	0.4524	
LS34	0.1152	0.0862		0.0000	0.0000		0.3501	0.0976	
LS39	0.0000	0.0000		0.5492	0.3889		0.4963	0.4048	
LS54	0.8321	0.7119		0.7758	0.4800		0.4887	0.0476	
AnacE4	0.8100	0.5484		0.6540	0.3889		0.5635	0.1429	
AnacC11	0.7243	0.7833		0.6372	0.4259		0.7735	0.6667	
AoxD234	0.8061	0.3387		0.8146	0.9815		0.7755	1.0000	
Average +/- S.D.	0.5704± 0.3549	0.4310± 0.3010	5.5714	0.6049± 0.2843	0.4760± 0.2989	6.2857	0.5634± 0.1577	0.4017± 0.3454	3.5714

Table 2. Expected heterozygosity (H_E), observed heterozygosity (H_O) and the mean number of alleles per locus.

The heterozygosity represents an important index that can give information about the diversity and even about the history of a population. The values of the heterozygosity can range from 0 (absence of heterozygosity) to 1 (significant number of alleles with the same frequency). Thus for the locus LS34 in *H. huso* and for LS39 in *A. ruthenus* we obtained a heterozygosity that is equal to 0, since a fixed allele was highlighted for these loci. The values of the average heterozygosity are correlated in direct proportion with genetic diversity.

The Hardy-Weinberg equilibrium was tested with Genepop v1.2 software. Except the p-value for LS19 and LS39 in *A. ruthenus*, the other loci showed significant departures from equilibrium. In generally, the deviation from the expected values might have several causes like the reduced size of the population, inbreeding or the presence of the null alleles which might lead to a false excess of homozygotes.

Our studies come to complete the data of evaluation and monitoring for the population of sturgeon from the Lower Danube. Thus, the data resulting from the monitoring of the YOYs (Young of the Year) born annually in the Romanian part of the Danube and evaluating the success of the natural recruitment have led to the hypothesis that in case of *H. huso* there is a significant generation born before 1990, which represents the basis of gene pool and which cyclic give birth to new generations with a high number of individuals [31].

5. Molecular studies for genetic diversity evaluation and phylogeny inferring in salmonids from Romania

The native salmonid species from Romania have been characterized only from a morphological point of view [28], but the studies based on molecular aspects are still at the beginning. Among the few studies that included the molecular analysis of salmonid fishes from Romania, the evaluation of the genetic differentiation of salmonids by PCR-RFLP technique [76] and the phylogenetic classification of Romanian salmonid species by using the 16SrRNA and 12SrRNA gene sequences [26] were performed.

Even if the salmonids are a well-studied group of fish, there are still a number of questions pending with regard to their phylogeny and evolution. So, despite the fact that a large number of studies based on both morphological [77] and molecular data [78-80] were performed, there are still different opinions concerning genus-level relationships [26]. Four salmonid species from Romania (*Salmo trutta fario*, *Salmo labrax*, *Salvelinus fontinalis* and *Thymallus thymallus*) were analyzed from molecular point of view using 16SrRNA and 12SrRNA markers with the purpose to position them within the *Salmonidae* family. The biological samples were collected from different rivers from Romania (Dambovita, Bratia, Gilau, Latorita, Cerna, and Nera) and from the Danube Delta. Fragments from 16SrRNA and 12SrRNA mitochondrial genes were amplified with specific primers and sequenced by using Sanger method, *dye terminator* variant. For a more complex phylogenetic evaluation beside the sequences determined from salmonid specimens from Romania, 14 salmonid and 1 osmerid sequences from GenBank were also included in the analysis.

Estimation of phylogenetic relationships was achieved using 16S and 12SrRNAs gene sequences and the concateneted data set, while three methodologies – maximum parsimony (MP), maximum likelihood (ML) and Neighbour-joining (NJ) implemented in PHYLIP v3.68 software were used for phylogenetic reconstructions, in order to compare the consistency of the results produced by different methods.

The phylogenetic analysis revealed that primitive salmonid species such as *Coregonus lavaraetus* and the representatives of genus *Thymallus*, *T. thymallus* and *T. articus* occupy basal divergence in the tree topology confirming that the *Coregoninae* and *Thymallinae* subfamilies arise from a common ancestry before *Salmoninae* (with the genera *Salmo, Oncorhynchus, Hucho, Brachymystax* and *Salvelinus*). Based on morphological and molecular data, *Coregoninae* and *Thymallinae* were thought to be the earliest branches within the *Salmonidae* family [80].

The species of the *Salmo* genus form a distinct clade, in which the Atlantic salmon, *Salmo salar* occupies a basal divergence. The data reveal a close relationship between *Salmo trutta fario* and the clade formed by sea trout *S. trutta trutta* and the Black Sea trout *S. labrax*. The resulting clade (*S. trutta trutta, S. labrax*) is not surprising, taking into consideration some characteristics of the life history and reproductive behavior of these species [8].

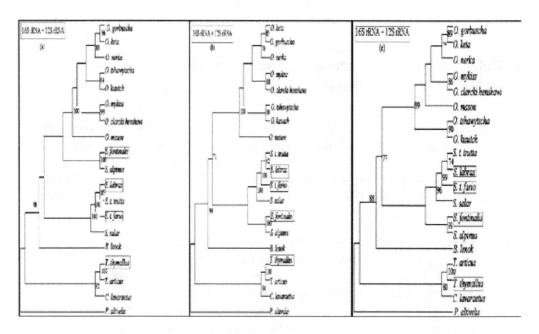

Figure 5. Majority with bootstrap support consensus trees for combined data (16SrRNA and 12SrRNA). (a) Combined data Neighbor Joining tree, distance model Kimura 2 Parameters, transition/transversion ratio 2.3; (b) combined data Maximum Parsimony tree; (c) combined data Maximum Likelihood tree [26].

The monophyly of *Salvelinus* was supported by 16SrRNA, 12SrRNA and combined data, but the position of the clade formed by (*S. alpinus, S. fontinallis*) in relationship with *Salmo* and *Onchorhynchus* is dependent on the molecular marker selected for the phylogenetic analysis. The trees resulted for 16SrRNA and 12SrRNA concatenated data has demonstrated a closer relationship between *Salvelinus* and *Oncorhynchus* than between *Salmo* and *Oncorhynchus*. The relationship between *Salvelinus* and *Oncorhynchus* was also supported by previous molecular phylogeny studies based on three other genes (GH1C, VIT and ND3) [80, 81].

The phylogenetic analysis using mitochondrial ribosomal genes as markers has allowed for the classification of salmonid species from Romania within the *Salmonidae* family. Thus, Romanian *S. trutta fario* and *S. labrax* are placed together within the *Salmo* genus. The *S. labrax*, endemic in the Black Sea, appears to be the sister taxa of the sea trout *S. trutta trutta* from the northwest of Europe (Atlantic coast) and Baltic Sea. The basal divergence in phylogenetic trees occupied by *T. thymallus*, a primitive species, is in agreement with the taxonomic and evolutionary data. Unfortunately, the position of the *Salvelinus* genus relative to the *Salmo* and *Oncorhynchus* genera remains controversial. The data reveal a possible sister taxon relationship between *Oncorhynchus* and *Salvelinus* despite the fact that morphological data support a closer relationship between *Salmo* and *Oncorhynchus*, thus confirming earlier findings of Crespi & Fulton [80] and Oakley & Phillips [81].

Unfortunately, salmonids' natural habitat is disrupted by a series of human activities such as poaching, dams, ballast exploitation and the construction of the micro hydro plants on the water streams from mountain areas. In this context, a preliminary study was directed to towards the assessment of the anthropic impact on the brown trout (*S. trutta fario*) populations from different rivers in Fagaras Mountain, including beside ecological aspects, molecular ones [82]. The area of Fagaras Mountain is the most representative in Romania for the brown trout populations, the tourism and recreational fishing being well known in this area, while the "brown trout of Fagaras" is preferred for alimentary consumption.

In the proposed study 102 individuals of brown trout from four populations from the Meridional Carpathians (Arpas, Ucea and Sambata rivers in the Northern versant of the Fagaras Mountain) were analyzed using mtDNA marker (D-loop) and nuclear markers (nine microsatellite loci - Str73, Str15, Str60, OmyFgt1, Ssa197, Ssa85, Strutta12, Str543, BS131).

The data obtained by sequencing of the complete mitochondrial control region (D-loop) were compared with similar sequences from GenBank. The phylogenetic tree resulted from analysis with MEGA v.5 contained, besides the sequences of D-loop in *S. trutta fario* individuals sampled in the Romanian rivers, complete D-loop sequences from other European lineages in order to observe the affiliation of analyzed Romanian individuals. The dendogram topology showed a classification of analyzed sequences in distinct monophiletic groups corresponding to each evolutive lineage and that the representatives of brown trout from Romania analyzed in the study were placed in the Danubian clade similar to other sequences selected from GenBank and belonging to Danubian lineage (Figure 6).

In Romania, Danubian lineage of brown trout is native and by founding the specific haplotypes for this lineage in our individuals, means that either restocking programs have not been done yet in the area or, if they have, the individuals involved were selected properly [82]. This type of approach can be really useful when the data about restocking and management of the rivers are poor or completely missing.

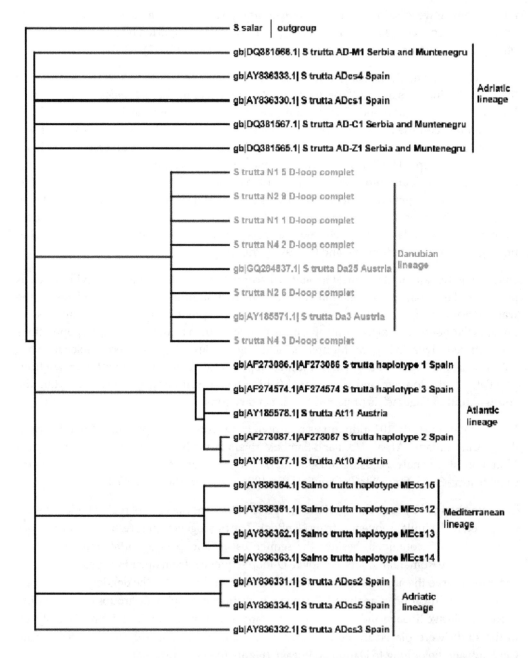

Figure 6. Phylogenetic tree obtained with MEGA v5 (NJ method, bootstrap value 1000 replications) illustrating the relationships between the evolutive lineages of brown trout. The haplotypes from Romania are placed in the Danubian lineage along with two sequences from GenBank belonging to the same lineage [82].

The genetic differentiation between the four population of brown trout analyzed in this study was evaluated by F-statistics, by using the indices F_{st} and $Gamma_{st}$ calculated with DNAsp software. F_{st} (fixation index) is a Wright statistic index that indicates the genetic variation between the populations and that can take values from 0 to 1. In generally, a higher value than 0.25 is correlated with a high genetic differentiation between populations. $Gamma_{st}$ is similar to F_{st}, making a correction for insufficient sampling. The values obtained for the four populations of brown trout indicate that there is a high differentiation between them from genetic point of view (Figure 7).

Population 1	Population 2	GammaSt	Fst
N1	N2	0.59009	0.72259
N1	N4	0.39013	0.55081
N1	N3	0.44519	0.58468
N2	N4	0.19504	0.30877
N2	N3	0.12098	0.17850
N4	N3	0.18104	0.28120

Figure 7. F_{st} and $Gamma_{st}$ values for four populations of brown trout from Romanian rivers [82].

The microsatellite data confirmed the previous results obtained regarding the genetic differentiation. Based on genotypic data resulted consequently microsatellites analysis a Factorial Correspondence Analysis (FCA) was performed by using GENETIX software. This type of analysis assumes the applying of a multidimensional method that permit the conversion of genotypic data obtained by the analysis of the nine microsatellite loci characteristic to each analyzed individual in points distributed in an X, Y, Z axis system. Thus, consequently to this analysis four different clusters were obtained corresponding to the four populations of brown trout, indicating a genetic differentiation between the analyzed populations (Figure 8).

The value of F_{is} index in the analyzed population indicates is positive for each population and indicates a light level of inbreeding within the population (Table 3).

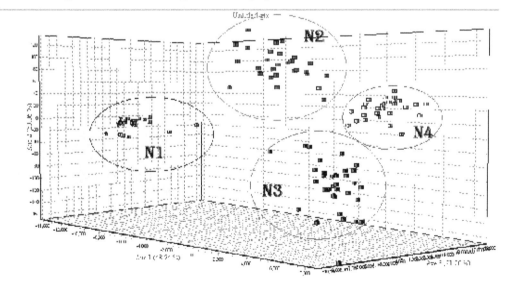

Figure 8. Factorial Correspondence Analysis for four populations of brown trout from Romanian rivers.

Locus	Population N1	Population N2	Population N3	Population N4
Str60	0.000	-0.250	0.477	-0.747
Str15	0.618	-0.190	NA	0.444
Str73	NA	NA	NA	NA
OmyFGT	0,143	-0.052	0.224	0.179
Ssa85	-0.500	NA	NA	0.505
Ssa197	0.191	0.254	0.017	0.094
Str543	0.020	0.911	-0.053	0.209
Strutta12	0.094	0.358	0.153	0.399
BS131	1.000	0.259	0.253	-0.127
Average	0.094	0.270	0.139	0.134

Table 3. F_{is} values calculated with FSTAT software. NA (not assigned).

Concluding, the analyzed populations are natives (Danubian lineage), differentiated by genetic point of view, but with a light level of inbreeding within population which might conduct in time to a reduced genetic diversity.

The studies are intent to be extended first to the Southern side of Fagaras Mountain, which is considered to be the most affected of the anthropic intervention, then to the main basins in the country populated with brown trout. The information resulted from these studies can be further used in programs of management and conservation, with accent on keeping unmodified the autochthonous species.

6. Conclusions

The studies developed in our country for analyzing in terms of genetic diversity species and populations of great scientific, ecological, economic and social importance are still at the beginning.

The molecular analyses of Lower Danube sturgeons were directed to species identification and assessing of genetic diversity. Regarding the first topic, a method based on microsatellite markers was set up as it was proven that the identification of species based only on morphology and mtDNA analysis can be misleading and do not serve to hybrid detection. The populations of sturgeon from Lower Danube appear to be fragile from genetic diversity point of view, but the studies should be extended in a higher number of individuals and using a more significant and informative set of molecular markers

The analysis of microsatellites in these species is complicated by the complexity of the genome, so a better selection of a much significant number of loci with disomic inheritance pattern is recommendable. The decision to initiate restocking programs in order to recover the natural sturgeon populations must take into account not only the social and economic aspects, but also the assessment of genetic diversity.

The studies on salmonid were focused on mainly on phylogenetic classification of species from Romanian fauna. The research aiming on genetic diversity analysis is still in an incipient phase and was directed towards the populations of brown trout from Fagaras Mountain, an important area for salmonid distribution in our country. The results of these studies based on mtDNA and microsatellites analysis showed that the studied populations are pure Danubian brown trout lineage and genetically distinct. In the future, for a better image of salmonids status in Romania the research would be extended in analysis populations from other salmonid species and the area of sampling would be enlarged to the entire Romanian Carpathian Arch.

7. Perspectives

As our studies are in the preliminary phase, in future we intend to extend them in a higher number of individuals/populations from both fish groups consider of being of highly importance for our country.

Also, we intend to characterize from genetic point of view populations/ stocks/ strains from aquaculture, as we consider that exploitation of genetic data is imperious necessary beside restocking and conservation, for genetic improvement in aquaculture.

Currently the microsatellites and mtDNA are considered to be classical markers. There is an increasing tendency of analyzing the nucleotide variation at the whole genome level by *Next Generation Sequencing (NGS)* techniques. This type of approach started to be applied for different fish species analysis, including sturgeons and salmonids. NGS variants as *RAD (Restriction Sites Associated DNA) sequencing* are used for identification and characterization of a complete panel of SNP markers consider to be extremely useful for genomic analyses at individual, population and species level. This type of markers appears to be highly informative and of real support for conservation programs as is the trend of passing from the conservation genetics to conservation genomics era. The identification by NGS of an extended panel of specific markers should be extremely useful for sturgeons from the Lower Danube in population management and conservation purposes. The analysis should be orientated toward *A. gueldenstaedtii* the most affected sturgeon species from the Danube. This is a polyploid species, presenting a complex genome structure, fact that makes very difficult the analysis using classical markers. Also, the setup of such complex panel of SNP markers would make possible the traceability of caviar in the sturgeon species from Romania.

Acknowledgements

This work was supported by the PN-II-PT-PCCA Project 116/2012 "Genetic evaluation and monitoring of molecular and biotechnological factors that influence productive performance of Danube sturgeon species bred in intensive recirculating systems". Dudu Andreea was supported by the European Social Funding through the Sectorial Operational Programme for Human Resources Development POSDRU/159/1.5/S/133391.

Author details

Andreea Dudu*, Sergiu Emil Georgescu and Marieta Costache

*Address all correspondence to: tn_andreea@yahoo.com

University of Bucharest, Department of Biochemistry and Molecular Biology, Splaiul Independentei 91-95, Bucharest, Romania

References

[1] Global Issues. http://www.globalissues.org/article/170/why-is-biodiversity-impor-tant-who-cares, accessed at 23 July 2014.

[2] Millennium Ecosystem Assessment, 2005. Ecosystems and Human Well-being: Biodi-versity Synthesis. World Resources Institute, Washington DC, http://www.millenniu-massessment.org/documents/document.356.aspx.pdf, accesed at 01 September 2014.

[3] Frankham R, Ballou JD, Briscoe DA. Introduction to Conservation Genetics. New York: Cambridge University Press; 2002.

[4] Wan QH, Wu H, Fujihara T, Fang SG. Which genetic marker for which conservation genetics issue? Electrophoresis 2004; 25:2165–2176.

[5] Soule ME, Wilcox BA. Conservation Biology. An Evolutionary–Ecological Perspec-tive. Massachusetts, USA: Sinauer Associates; 1980.

[6] FishBase http://www.fishbase.org/search.php, accessed at 23 July 2014.

[7] The IUCN Red List of Threatened Species http://www.iucnredlist.org/about/summa-ry-statistics#Tables_1_2 accessed at 23 July 2014.

[8] Otel V. Atlasul pestilor din Rezervatia Biosferei Delta Dunarii. Tulcea, Romania: Edi-tura Centrul de Informare Tehnologica Delta Dunarii; 2007.

[9] Garrido-Ramos MA, Robles F, de la Herrán R, Martínez-Espín E, Lorente JA, Ruiz-Rejón C, Ruiz-Rejón M. Analysis of Mitochondrial and Nuclear DNA Markers in Old Museum Sturgeons Yield Insights About the Species Existing in Western Europe: *A. sturio, A. naccarii* and *A. oxyrinchus.* In Carmona R, Domezain A, Gallego MG, Her-nando JA, Rodríguez F, Ruiz-Rejón M. (Ed) Biology, Conservation and Sustainable Development of Sturgeons. Fish and Fisheries Series, 1st Edition, Amsterdam, Neth-erlands: Springer 2009. p25–50.

[10] Bacalbasa-Dobrovici N. Endangered migratory sturgeons of the Lower Danube River and its Delta. In: Balon EK (Ed) Envinronmental Biology of Fishes 1997, 48(1–4): 201-207.

[11] Kynard B, Zhuang P, Zhang T, Zhang L. Ontogenetic behavior and migration of Volga River Russian sturgeon, *Acipenser gueldenstaedtii*, with a note on adaptive significance of body color. Envinronmental Biology of Fishes 2002; 65:411–421.

[12] Navodaru I, Staras M, Banks R. Management of sturgeon stocks of the Lower Danube River system. In: The Delta`s: State-of art protection and management. Conference Proceedings, 26-31 July 1999; Tulcea, Romania.

[13] Suciu R. Sturgeons of the NW Black Sea and Lower Danube River countries. At: International Expert Workshop on CITES Non-Detriment Findings; 17-22 November 2008, Cancun, Mexico.

[14] Ludwig A, Lippold S, Debus L, Reinartz R. First evidence of hybridization between endangered sterlets (*Acipenser ruthenus*) and exotic Siberian sturgeons (*Acipenser baerii*) in the Danube River. Biological Invasions 2009; 11:753–760.

[15] Dudu A, Suciu R, Paraschiv M., Georgescu SE, Costache M, Berrebi P. Nuclear Markers of Danube Sturgeons Hybridization, International Journal of Molecular Sciences 2011; 12(10):6796-6809.

[16] Ludwig A. Identification of Acipenseriformes species in trade. Journal of Applied Ichthyology 2008; 24(S1):2–19.

[17] Fain S, Straughan D, Hamlin B, Hoesch R, LeMay J. Forensic genetic identification of sturgeon caviars traveling in world trade. Conservation Genetics 2013; 14(4):855-874.

[18] Boscari E, Barmintseva A, Pujolar JM, Doukakis P, Mugue N, Congiu L. Species and hybrid identification of sturgeon caviar: a new molecular approach to detect illegal trade. Molecular Ecology Resources 2014; 14:489–498.

[19] Nelson JS. Fishes of the World, 4th Edition. New York: John Wiley & Sons; 2006.

[20] Romero A. Salmoniformes (Salmons), In: Hutchins M, Thoney DA, Loiselle PV, Schlager N. (ed) Grzimek's Animal Life Encyclopedia, 2nd Edition, Vol. 4, Fishes. Farmington Hilla: Gale Group; 2003. p. 405-420.

[21] Kottelat M, Freyhof J. Handbook of European freshwater fishes. Berlin, Germany: Freyhof, 2007; p. 429-430.

[22] Bernatchez L. The evolutionary history of brown trout (*Salmo trutta* L.) inferred from phylogeographic, nested clade, and mismatch analysis of mitocondrial DNA variation. Evolution 2001; 55:351-379.

[23] Meraner A, Baric S, Pelster B, Dalla Via J. Trout (*Salmo trutta*) mitochondrial DNA polymorphism in the centre of the marble trout distribution area. Hydrobiologia 2007, 579:337–349.

[24] Giuffra E, Bernatchez L, Guyomard R. Mitochondrial control region and protein coding genes sequence variation among phenotypic forms of brown trout *Salmo trutta*
from northern Italy. Molecular Ecology 1994; 3:161–171.

[25] Osinov AG, Bernatchez L. Atlantic and Danubian phylogenetic groupings of brown
trout *Salmo trutta* complex: genetic divergence, evolution, and conservation. Journal
of Ichthyology 1996; 36:723–746.

[26] Dudu A, Georgescu SE, Popa O, Dinischiotu A, Costache M. Mitochondrial 16s and
12s rRNA sequence analysis in four salmonid species from Romania. Acta Zoologica
Academiae Scientiarum Hungaricae 2011, 57(3):233-246.

[27] Hindar K, Jonsson B, Ryman N, Staahl G. Genetic relationships among landlocked,
resident, and anadromous brown trout, Salmo trutta L. Heredity 1991; 66:83-91.

[28] Bănărescu P. Fauna Republicii Populare Romîne. Piscies-Osteichthyes (Peşti ganoizi
şi osoşi) Vol. XIII. Bucureşti: Editura Academiei R.P.R.; 1964.

[29] Bănăduc D. The Hucho hucho (Linnaeus, 1758), (Salmoniformes, Salmonidae), species monitoring in the Viseu River (Maramures, Romania), Transylvanian Review of
Systematical and Ecological Research 2008, 5:183–188.

[30] Ciolac A, Patriche N. Biological aspects of main marine migratory sturgeons in Romanian Danube River. Migration of fishes in Romanian Danube River. Applied Ecology and Environmental Research 2005; 3(2):101-106.

[31] Paraschiv M, Suciu R, Suciu M. Present state of sturgeon stocks in the Lower Danube
River, Romania. Proceedings 36th International Conference of IAD, Austrian Commit-
tee Danube Research / IAD, Vienna; 2006.

[32] Lenhardt M, Cakic P, Kolarevic J, Gacic Z. Morphometric recognition of two morphs
in sterlet (*Acipenser ruthenus*) population induced by different reproductive behaviour. Fish as models of behavior. The Fisheries Society of the British Isles. Annual International Symposium, University of East Anglia, Norwich, England; 30 June – 4
July 2003.

[33] http://www.merriam-webster.com, accesed at 01 September 2014.

[34] Liu ZJ, Cordes JF. DNA marker technologies and their applications in aquaculture
genetics. Aquaculture 2004, 238:1-37.

[35] O'Brien SJ. Molecular genome mapping: lessons and prospects. Current Opinion in
Genetics and Development 1991, 1:105–111.

[36] Houston RD, Haley CS, Hamilton A, Guy DR, Tinch AE, Taggart JB, McAndrew BJ,
Bishop SC. Major quantitative trait loci affect resistance to infectious pancreatic ne-
crosis in Atlantic salmon (*Salmo salar*). Genetics 2008, 178:1109-1115.

[37] Baranski M, Moen T, Vage DI. Mapping of quantitative trait loci for flesh colour and
growth traits in Atlantic salmon (*Salmo salar*). GenetIcs Selection Evolution 2010,
42:17.

[38] Norman JD, Robinson M, Glebe B, Ferguson MM, Danzmann RG. Genomic arrangement of salinity tolerance QTLs in salmonids: a comparative analysis of Atlantic salmon (*Salmo salar*) with Arctic charr (*Salvelinus alpinus*) and rainbow trout. BMC Genomics 2012, 13:420.

[39] Gutierrez AP, Lubieniecki KP, Fukui S, Withler RE, Swift B, Davidson WS. Detection of Quantitative Trait Loci (QTL) related to grilsing and late sexual maturation in Atlantic Salmon (*Salmo salar*). Marine Biotechnology 2014; 16(1):103-109.

[40] Szalanski AL, Bischof R, Holland R. Mitochondrial DNA variation in pallid and shovelnose sturgeon. Transactions Nebraskan Academy of Sciences (Affil. Soc.) 2001; 26: 19-21.

[41] Waldman JR, Grunwald C, Stabile J. Impacts of life history and biogeography on the genetic stock structure of Atlantic sturgeon Acipenser oxyrinchus oxyrinchus, Gulf sturgeon A-oxyrinchus desotoi, and shortnose sturgeon A. brevirostrum. Journal of Applied Ichthyology 2002; 18:509-518.

[42] Wolf C, Hübner P, Lüthy J. Differentiation of sturgeon species by PCR-RFLP. Food Research International 1999; 32:699-705.

[43] Panagiotopoulou H, Baca M, Popovic D, Weglenski P, Stankovic A. A PCR-RFLP based test for distinguishing European and Atlantic sturgeons. Journal of Applied Ichthyology 2014, 30:14–17.

[44] Ludwig A, Debus L, Jenneckens I. A molecular approach for trading control of black caviar. International Review of Hydrobiology 2002; 87:661-674.

[45] Ludwig A, Congiu L, Pitra C, Fickel J. Nonconcordant evolutionary history of maternal and paternal lineages in Adriatic sturgeon. Molecular Ecology 2003; 12: 3253-3264.

[46] Machordom A, Suarez J, Almodovar A, Bautista JM. Mitochondrial haplotype variation and phylogeography of Iberian brown trout populations. Molecular Ecology 2000; 9:1325–1338.

[47] Lucentini L, Palomba A, Gigliarelli L, Lancioni H, Viali P, Panara F. Genetic characterization of a putative indigenous brown trout (*Samo trutta fario*) population in the secondary stream of the Nera River Basin (Central Italy) assessed by means of three molecular markers. Italian Journal of Zoology 2006; 73:263–273.

[48] Oleinik AG, Skurikhina LA, Brykov VA. Genetic Differentiation of Three Sympatric Charr Species from the Genus Salvelinus Inferred from PCR-RFLP Analysis of Mitochondrial DNA. Russian Journal of Genetics 2003; 39(8):924.

[49] Koskinen MT, Ranta E, Piironen J, Veselov A, Titov S, Haugen TO, Nilsson J, Carlstein M, Primmer CR. Genetic lineages and postglacial colonization of grayling (*Thymallus thymallus*, Salmonidae) in Europe, as revealed by mitochondrial DNA analyses. Molecular Ecology 2000; 9(10):1609-24.

[50] Tautz D. Hipervariability of simple sequences as a general source for polymorphic DNA markers. Nucleic Acids Research 1989; 17: 6463-6471.

[51] Wright JM. DNA fingerprinting in fishes. In Hochachka PW, Mommsen T. (ed.) Biochemistry and Biology of Fishes, Vol. 2, Amsterdam:Elsevier, 1993; p.57-91.

[52] Tóth G, Gáspári Z, Jurka J. Microsatellites in different eukaryotic genomes: Survey and analysis. Genome Research 2000; 10(7):967-981.

[53] Botstein D, White RL, Skolnick M, Davis RW. Construction of a genetic linkage map in man using restriction fragment length polymorphisms. American Journal of Human Genetics 1980; 32:314–331.

[54] Estoup A, Angers B. Microsatellites and minisatellites for molecular ecology: theoretical and empirical considerations. In: Carvalho G. (ed.) Advances in molecular ecology. Amsterdam: IOS Press, 1998; p. 55–86.

[55] King TL, Lubinski BA, Spidle AP. Microsatellite DNA variation in Atlantic sturgeon (Acipenser oxyrinchus oxyrinchus) and cross-species amplification in the Acipenseridae. Conservation Genetics 2001; 2: 103-119.

[56] Welsh A., May B. Development and standardization of disomic microsatellite loci for lake sturgeon genetic studies. Journal of Applied Ichthyology 2006; 22:337–44.

[57] Welsh AB, Blumberg M, May B. Identification of microsatellite loci in lake sturgeon, Acipenser fulvescens, and their variability in green sturgeon, A. medirostris. Molecular Ecology Notes 2003; 3: 47–55.

[58] Forlani A, Fontana F, Congiu L. Isolation of microsatellite loci from the endemic and endangered Adriatic sturgeon (Acipenser naccarii). Conservation Genetics 2008; 9: 461–463.

[59] Boscari E, Barbisan F, Congiu L. Inheritance pattern of microsatellite loci in the polyploid Adriatic sturgeon (*Acipenser naccarii*). Aquaculture 2011; 321(3–4):223–229.

[60] Moghim M, Pourkazemi M, Tan S, Siraj S, Panandam J, Kor D. Development of disomic single-locus DNA microsatellite markers for Persian sturgeon (*Acipenser persicus*) from the Caspian Sea. Iranian Journal of Fisheries Sciences 2013; 12 (2):389-397.

[61] Norouzi M, Pourkazemi M. Genetic structure of Caspian populations of stellate sturgeon, *Acipenser stellatus* (Pallas, 1771), using microsatellite markers. International Aquatic Research 2009; 1:61-65.

[62] Timoshkina N, Barmintseva AE, Usatov AV, Mugue NS. Intraspecific genetic polymorphism of Russian sturgeon *Acipencer gueldenstaedtii*. Russian Journal of Genetics 2009, 45:1098–1107.

[63] Koljonen ML, Tähtinen J, Säisä M, Koskiniemi J. Maintenance of genetic diversity of Atlantic salmon (*Salmo salar*) by captive breeding programmes and the geographic distribution of microsatellite variation. Aquaculture 2003; 212, 69–92.

[64] Apostolidis AP, Madeira MJ, Hansen M, Machordom A. Genetic structure and demographic history of brown trout (*Salmo trutta*) populations from the southern Balkans. Freshwater Biology 2008; 53:1555–1566.

[65] Nilsson J, Ostergren J, Lundqvist H, Carlsson U. Genetic assessment of Atlantic salmon Salmo salar and sea trout Salmo trutta stocking in a Baltic Sea river. Journal of Fish Biology 2008; 73:1201–1215.

[66] Pourkazemi M, Skibinski DO, Beardmore JA. Application of mtDNAd-loop region for the study of Russian sturgeon population structure from Iranian coastline of the Caspian sea. Journal of Applied Ichthyology 1999; 15:23-28.

[67] Zhang SM, Deng H, Wang DQ, Zhang YP, Wu QJ. Mitochondrial DNA Length variation and heteroplasmy in Chinese sturgeon (*Acipenser sinensis*). Acta Genetica Sinica 1999; 26, 489–496.

[68] Doukakis P, Birstein VJ, DeSalle R. Molecular genetic analysis among subspecies of two Eurasian sturgeon species, Acipenser baerii and A. stellatus. Molecular Ecology 1999; 12:S117–S129.

[69] Asplund T, Veselov A, Rimmer C, Bakhmet I, Potutkin A, Titov S, Zubchenko A, Studenov I, Kaluzchin S, Lumme J. Geographical structure and postglacial history of mtDNA haplotypes variation in Atlantic salmon (*Salmo salar* L.) among rivers of the White and Barents Sea basins. Annales Zoologici Fennici 2004, 41:465–475.

[70] Laikre L, Järvi T, Johansson L, Palm S, Rubin JF, Glimsater CE, Landergren P, Ryman N. Spatial and temporal population structure of sea trout at the Island of Gotland, Sweden, delineated from mitochondrial DNA. Journal of Fish Biology 2002; 60:49–57.

[71] Frankel OH, Soulé ME. Conservation and evolution. Cambridge University Press, 1981; Cambridge, UK.

[72] Vasil'ev VP. Mechanisms of Polyploid Evolution in Fish: Polyploidy in Sturgeons. In Carmona R, Domezain A, Gallego MG, Hernando JA, Rodríguez F, Ruiz-Rejón M. (Ed) Biology, Conservation and Sustainable Development of Sturgeons. Fish and Fisheries Series, 1st Edition, Amsterdam, Netherlands: Springer 2009. p97–117.

[73] Birstein VJ, Bemis WE. How Many Species are there within the Genus Acipenser? In Sturgeon Biodiversity and Conservation, 1st edition; Dordrecht, Netherlands: Kluiwer Academic Publishers, 1997. p157–163.

[74] Ludwig A, Belfiore NM, Pitra C, Svirsky V, Jenneckens I. Genome duplication events and functional reduction of ploidy levels in sturgeon (*Acipenser, Huso* and *Scaphirhynchus*). Genetics 2001, 158:1203–1215.

[75] Jenneckens I, Meyer JN, Hörstgen-Schwark G, May B. A fixed allele at microsatellite LS-39 is characteristic for the black caviar producer Acipenser stellatus. Journal of Applied Ichthyology 2001; 17:39-42.

[76] Dudu A, Georgescu SE, Dinischiotu A, Costache M. PCR-RFLP method to identify fish species of economic importance, Archiva Zootechnica 2010; 13(1):53-59.

[77] Stearley RF, Smith GR. Phylogeny of the Pacific trouts and salmons (*Oncorhynchus*) and genera of the family *Salmonidae*. Transaction of American Fisheries Society 1993; 122:1–33.

[78] Kitano T, Matsuoka N, Saitou N. Phylogenetic Relationship of the genus *Oncorhynchus* species inferred from nuclear and mitochondrial markers, Genes Genetics and Systematics 1997; 72:25–34.

[79] Oohara I, Sawano K, Okazaki T. Mitochondrial DNA Sequence Analysis of the Masu Salmon—Phylogeny in the Genus Oncorhynchus. Molecular Phylogenetics and Evolution 1997; 7:71–78.

[80] Crespi BJ, Fulton MJ. Molecular systematics of *Salmonidae*: combined nuclear data yields a robust phylogeny. Molecular Phylogenetics and Evolution 2004; 31:658–679.

[81] Oakley TH, Phillips RB. Phylogeny of Salmonine Fishes Based on Growth Hormone Introns: Atlantic (*Salmo*) and Pacific (*Oncorhynchus*) Salmon Are Not Sister Taxa. Molecular Phylogenetics and Evolution 1999; 11(3):381–393.

[82] Popa GO, Khalaf M, Dudu A, Curtean-Bănăduc A, Bănăduc D, Georgescu SE, Costache M. Brown trout's populations genetic diversity using mitochondrial markers in relatively similar geographical and ecological conditions – A Carpathian case study, Transylvanian Review of Systematical and Ecological Research 2013, 15(2):125-132.

6

Genetic Diversity in Bananas and Plantains (*Musa* spp.)

G. Manzo-Sánchez, M.T. Buenrostro-Nava,
S. Guzmán-González, M. Orozco-Santos,
Muhammad Youssef and
Rosa Maria Escobedo-Gracia Medrano

1. Introduction

Bananas and plantains belong to the family *Musaceae* and are cultivated throughout the humid tropics and sub-tropics. This crop is perennial with a faster relative growth rate compared to other fruit crops, while producing fruit all year round. Because of their nutritional value, bananas and plantains are considered the fourth most important crop worldwide after rice, wheat and corn. In many countries of Africa, bananas are considered an important part of the diet; the population in Uganda consumes per capita an average of 191 kg per year [1]. This crop also represents an important source of income for many rural families that work directly or indirectly in this industry. The edible *Musa* spp. originate from two wild species, *Musa acuminata* Colla and *M. balbisiana* Colla, with the A and B genomes, respectively, as well as their hybrids and polyploids.

The genus *Musa* is of great importance worldwide due to the commercial and nutritional value of cultivated varieties. Morphological data have suggested that *Musa* is diverse, with well-defined characteristics giving a number of indicators of the genome constitution. However, phenotyping for many physiological characteristics, including biotic and abiotic stress tolerance, particularly under controlled, contained and reproducible conditions, is difficult because of the size of the plants and their long life cycle.

DNA marker technologies have been widely used in banana genetics and diversity analysis, e.g., in taxonomy, cultivar true-to-type assessment and genetic linkage map development. Currently, proteomic analysis is giving rise to new trends in genetic diversity and plant system biology analyses. These approaches will yield detailed insights into the *Musa* genome and provide important genetic data for *Musa* breeders. In this chapter, we discuss the contribution of different DNA and protein-based markers to understanding the genetic diversity of the *Musaceae* family.

2. *Musa* genome

The *Musaceae* family genome is complex, and there are four genomes present, corresponding to the genetic constitutions of the four wild *Musa* species, i.e., *M. acuminata* (A-genome, 2n=2x=22), *M. balbisiana* (B-genome, 2n=2x=22), *M. schizocarpa* (S genome, 2n=2x=22) and *M. textillis* (T genome, 2n=2x=20). Modern bananas and plantain varieties are polyploids, and have a complex genetic structure; because most cultivated bananas are parthenocarpic (generally seedless) they have to be vegetatively propagated [2]. The traits of inflorescence in some accessions of *Musa* spp. are presented in Figure 1.

With the advent of modern DNA sequencing technologies and powerful bioinformatics tools, the sequencing and assembly of genomes for economically important crops and their relatives is becoming more common [3]. Knowing and understanding the genetic make-up of these crops represents a great opportunity to not only elucidate the function of genes of interest, but also to detect regions in the genome that could present polymorphism associated with agronomic traits [4]. This genetic variability is extremely valuable in plant breeding programmes where the selection of individuals with desirable characteristics is carried out.

Significant progress has being made towards gaining a better understanding of the *Musa* genome. Recently, [5] published the first draft of a 523-megabase *M. acuminata* genome and showed 36,542 protein-coding gene models. Transposable elements accounted for almost 50 % of the genome. More recently, a draft of the *M. balbisiana* genome sequence was published [6]. The data in that study showed a great divergence between the A and B genome, which is useful in terms of increasing genetic diversity. The *M. balbisiana* genome was shown to be 79 % smaller than that of *M. acuminata*, but with a highly similar number of predicted functional gene sequences: 36,638. Genomic information sheds light on polymorphisms that can be used in plant breeding programmes, such as the use of single nucleotide polymorphism (SNPs) where there is a high degree of heterozygosity between the A and B genome of one in every 55.9 base pairs. SNP polymorphisms have been successfully used with *Musa* spp. for mapping the genes *phytoene synthase* and *lycopene β-cyclase*, both of which are involved in β-carotene biosynthesis. Genomic information will bring about a strong and significant improvement in the use of DNA-based molecular markers, and make banana and plantain breeding programmes more efficient in improving the crop's traits.

A large dataset of genomic information is publicly available through different databases, such as The Banana Genome Hub (http://banana-genome.cirad.fr/home), resources from the Global Musa Genomics Consortium (GMGC; http://musagenomics.org/), and the National Center for Biotechnology Information (NCBI; http://www.ncbi.nlm.nih.gov/), among many others.

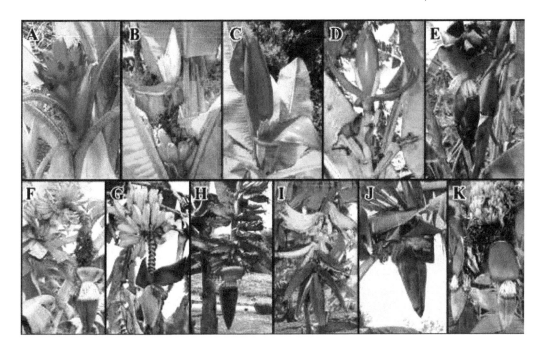

Figure 1. *Musa* accessions in a germplasm collection at Yucatan, Mexico. A; *Musa coccinea*, B: *M. velutina*, C: *M. laterita*, D: *M. beccarii*, E: *M. textiles*, F: *M. acuminata*, G: *M. balbisiana*, H: dessert banana (AAA), I: plantain (AAB), J: tetraploid hybrid (AAAB) and K: cooking banana (ABB) with two inflorescences. (Accessions as described in references [7, 8].)

3. DNA markers

In *Musa* breeding programmes, the use of morphological and cytogenetic markers has played an important role in identifying genes that control discrete traits with simple Mendelian inheritance, and has helped to estimate genome size and diversity [9]. However, a more detailed insight into the genomic polymorphism associated with agronomic traits is required.

Molecular DNA-based markers are powerful tools for gaining insights into individual genetic characteristics, and for determining allele frequency. DNA markers were developed first for humans, then applied in plants, and subsequently for the analysis of the banana genome (Figure 2). This feature allows plant breeders to select only those individuals with desirable characteristics and significantly reduce the selection time. In this chapter, we discuss the development and applications of molecular marker technology to improve some of the commercially available *Musa* cultivars and to assess their genetic diversity. The more important advantages and disadvantages of some DNA markers are presented in Table 1.

3.1. Restriction Fragment Length Polymorphism (RFLP)

RFLPs markers are widely used to detect variations in DNA fragment length banding patterns of electrophoresed restriction digests of DNA samples [10]. These variations are mainly due to the presence of a restriction enzyme cleavage site at one site in the genome of one individual, and the absence of the site in another individual. It can also detect changes in fragment size due to insertions or deletions between the restriction fragments. RFLP is a codominant marker, meaning that it is able to distinguish between homozygotes and heterozygotes. RFLP is robust, easily transferred between laboratories, and requires no prior sequence information for its use.

RFLPs have been found to be useful in *Musa* for constructing genetic linkage maps, characterizing germplasm, phylogenetic analysis [11-15] and analysis of variation in the chloroplast genome [16, 17], and most recently have been linked to polymorphisms in resistance gene analogues [18]. These markers were useful for detecting genetic variations in Indian wild *Musa balbisiana* populations associated with morphotaxonomic characterization clustering of most of the test types of bananas. However, they fail for some specific clusters [19], suggesting the need for more specific types of markers.

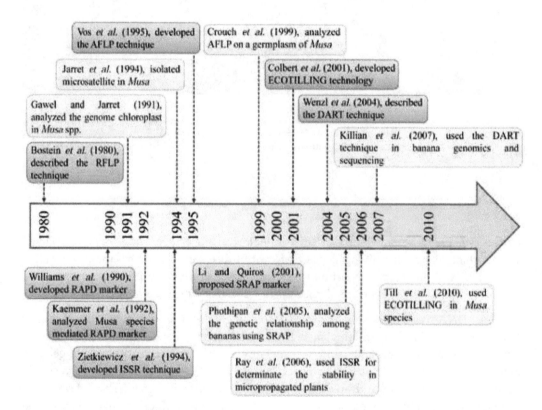

Figure 2. Schematic representation describing the development of DNA marker techniques and the first report on studies of the genetic diversity in *Musa* species.

RFLPs make locus-specific estimations of conserved synteny possible; however, some of the disadvantages are that it is expensive to develop, requires large amounts of DNA, is not possible to automate, unlike other DNA markers [AFLP, diversity array technology (DArT), or variable number tandem repeats (VNTR)], needs a suitable probe library, may require radioactive labelling, is laborious and time consuming. The relatively high cost and technically demanding nature of this technique make it inappropriate for routine breeding applications [20]. The use of more specific, PCR-based types of markers overcomes most of the disadvantages associated with RFLPs.

3.2. Random Amplified Polymorphic DNA (RAPD)

With the development of the polymerase chain reaction (PCR) technique, amplifying specific regions of an individual genome became possible, and identifying polymorphisms is made more precise by detecting small nucleotide changes compared to RFLPs. Random amplified polymorphic DNA (RAPD) depends on the PCR and is used as a very fast way to obtain information about genetic variation with a relatively low cost [21]. Some other characteristics are the fact that no prior knowledge of the genome sequence is required; low amounts of DNA template are used; and the advantage of technical simplicity. RAPD assays have proven to be powerful and efficient means of assisting introgression and backcross breeding [22]. But reproducibility is sometimes limited, and reliability depends on the skills of the operator, which is a dominant feature of the marker system.

RAPD has been widely used to distinguish diverse *Musa* germplasms [19, 23-26], for identification of duplications among accessions in tissue culture germplasm banks and somaclonal variation [27-29], and differentiation of irradiated banana genotypes [30, 31]. In addition, a molecular linkage map has also been developed using a variety of marker systems including RAPD [32]. Specific RAPD markers for the A and B genome of *Musa* have been identified [33, 34] and full-sib hybrids in plantain breeding populations [35]. These reports clearly demonstrate the potential value of this technique for germplasm characterization and cultivar identification, but give little insight into the value of the assay for molecular breeding. Despite the criticism of the technique, it is still being reported in recent publications [36-38].

Kaemmer [23] was first to report the use of RAPDs for fingerprinting of wild species and cultivars of banana (*Musa* spp.) by using simple sequence repeats (SSRs) as primers labelled with ^{a32}P. Fingerprinting analysis detected enough genetic variation to help discriminate between most of the 15 banana clones tested. In this study, polymorphic bands were unique to most of the clones tested and helped to distinguish between most of the wild type *M. balbisiana* (BB) and the *M. acuminata* (AAA) clones, as well as their hybrids for plantains (AAB) and cooking bananas (ABB); however, this technique generally fails to explain some of the variation within the species. Later, [24] used RAPD markers by adopting a set of nine primers that were shorter (N_{10}) than the SSRs (N_{16}) reported by [23], but the authors were able to find enough polymorphism that was unique to each of the nine genotypes representing the AA, AAA, AAB, ABB and BB genomes, and multivariate analysis showed a strong correlation between the polymorphism obtained and the morphological characters used to classify each of the groups.

The use of RAPDs was reported to identify 57 cultivars by using 60 10-mer random primers, where only 49 primers gave consistent results, and the primer OPC-15 ($^{5'}$GACGGATCAG$^{-3'}$) helped to distinguish 55 of the cultivars by producing 24 bands of all tested primers [25]. These markers failed to properly characterize the clones that Gros Michel and Venkel had previously thought belonged to the *Acuminata* group (AAA). However, chloroplast polymorphism was shown to be identical to *M. balbisiana*. This showed the potential of using RAPDs markers for proper cultivar identification and germplasm classification. Similar applications, but with a variant, were reported later by [17], who used a single primer ($^{5'}$TATAGTTAC-CAAGTGGTGGGGG$^{3'}$) designed from a human *Alu* sequence. This sequence is a member of well-conserved short interspersed nuclear elements in the primate's genome. Its use in fingerprinting the *Musa* genome produced bands that ranged between 300 bp and 3 kb, indicating that *Alu* related sequences are inverted repeats that are relatively close to each other.

3.3. Variable Number Tandem Repeats (VNTR)

VNTR are generated by highly specific PCR amplification and, therefore, should not suffer from the reproducibility problems experienced with RAPD analysis. VNTR are regions of short, tandemly repeated DNA motifs (generally less than or equal to 4 bp), with an overall length in the order of tens of base pairs [39]. VNTR have been reported to be highly abundant and randomly dispersed throughout the genomes of many plant species. Variation in the number of times the motif is repeated is thought to arise through slippage errors during DNA replication.

Furthermore, the isolation of VNTR is becoming increasingly routine with the availability of automated DNA sequencing facilities, along with improved techniques for the construction of genomic libraries enriched for VNTR and improved techniques for the screening of appropriate clones [40], bacterial artificial chromosome (BAC) end-sequences and, recently, the availability of genome sequencing facilities attributed to the discovery of VNTRs [41].

The development and utilization of VNTR in *Musa* research was first reported by [35, 40, 42] and [35]. VNTR have been considered optimum markers in other systems due to their abundance, polymorphism and reliability. VNTR analysis has been shown to detect a high level of polymorphism between individuals of *Musa* breeding populations [35], and used for development of a linkage map [18]. Nevertheless, several hundred VNTR markers have been generated in *Musa* [40, 42, 43]. New VNTR loci were discovered in the *M. acuminata* Calcutta 4 BAC end-sequence [41] from five fully sequenced consensi datasets, with validation for polymorphism conducted on genotypes contrasting in host plant resistance to Sigatoka disease [44]. Recently, [45] used a transcriptome database to design primers that were able to distinguish 32 VNTR and 119 target region amplified polymorphism (TRAP) alleles in 14 diploid *Musa* accessions.

3.4. Inter Simple Sequence Repeats (ISSR)

The ISSR technique developed by [46] does not require the knowledge of flanking sequences and has wide applications for all organisms, regardless of the availability of information about their genome sequence. They have also proved to be simple, fast, cost effective and versatile sets of markers for repeatable amplification of DNA sequences using single primers. As for the disadvantages, the homology of the bands is uncertain, and because they are dominant markers, they do not allow the calculation of certain parameters, requiring that heterozygous be distinguished from homozygous dominance.

ISSR and RAPD were used in determinate genetic stability of three economically important micropropagated banana (*Musa* spp.) cultivars. The results showed that ISSR detected more polymorphism than RAPD [27]. Similarly, ISSR were used for detection of genetic uniformity of micropropagated plantlets [47] and for screening *in vitro* mutagenesis and variance [37]. Another study reported the use of ISSR to assess the genetic diversity and classification of 27 wild banana accessions collected in Guangxi, China. The results showed that the collected germplasm was derived from diverse origins and evolutionary paths of banana in Guangxi [48]. ISSR were employed for molecular assessment of genetic identity and genetic stability in banana cultivars [49]. Recently, ISSR were used to analyse the pattern of genetic variation and differentiation in 32 individuals along with two reference samples of wild *Musa*, which corresponded to three populations across the biodiversity-rich hot-spot of the southern Western Ghats of India [50].

3.5. Sequence-Related Amplified Polymorphism (SRAP)

The sequence-related amplified polymorphism (SRAP) technique is a simple and efficient marker system that can be adapted for a variety of purposes, including map construction, gene tagging, genomic and cDNA fingerprinting, and map-based cloning. It has several advantages over other molecular marker systems, such as simplicity, reasonable throughput rate, disclosure of numerous codominant markers, ease of isolation of polymorphic bands for sequencing, and most importantly, the targeting of open reading frames (ORFs) [51].

The SRAP marker has been adopted recently for the assessment of genetic diversity and relationships in *Musa*. In this regard, the study of Thailand's wild landraces and cultivars of *M. acuminata* (A genome), *M. balbisiana* (B genome) and plantains using SRAP and RAPD has shown that the former marker was more efficient for detecting differences among closer cultivars in the same group, and the BB banana accessions were clustered separately from the AA banana accessions [52].

In another study, SRAP and AFLP were used to study 40 *Musa* accessions in a core collection being established in Mexico, which includes commercial cultivars and wild species of interest for genetic enhancement [7]. In addition to its practical simplicity, SRAP exhibited approximately three times more specific and unique bands than AFLP. Furthermore, SRAP was demonstrated to be a proficient tool for discriminating among *M. acuminata*, *M. balbisiana* and *M. schizocarpa* in the *Musa* section, as well as between plantains and cooking bananas within triploid cultivars. The six dessert banana cultivars used clustered according to their subgroup, i.e., Cavendish, Ibota and Gros Michel. Moreover, unique and specific bands were clearly recognized for each of seven subspecies of the *acuminata* complex, i.e., *microcarpa*, *malaccensis*, *zebrina*, *banksii*, *truncata* and *burmanica-burmanicoides* [7].

Moreover, the study of genetic relationships among some banana cultivars from China analysed by SRAP showed a correlation between the cultivars and their region of origin; the cultivars closely clustered into two major clusters according to their genome composition.

Likewise, the genetic data generated by the SRAP marker were reliable in respect to the morphology and agronomic trait classification, indicating the efficiency of SRAP for estimating genetic similarity among banana cultivars and providing a scientific basis for banana genetic and breeding research [53]. More recently, the fluorescently labelled SRAP molecular marker system was used to characterize the genetic variability within 71 accessions of a core collection, including wild species and cultivars of different subgroups [8], which complements previous work from the same collection [7].

The fluorescent SRAP marker information shows that *M. acuminata* subspecies *errans* was gathered with *banksii* in one cluster, while *malaccensis* was separated in a single sub-cluster. This study also found that Pisang Batu, which is known as diploid BB, was clustered with AB hybrids, i.e., Safet Velchi, Kunnan and Kamaramasenge, which supports the previous finding [54], and that *Pisang Batu* possesses a nucleotide polymorphism pattern of AA. Thus, the study suggested that *Pisang Batu* is a mislabelled accession of AAB or AB. Moreover, the SRAP marker system was found to be useful in identifying closely related accessions in the genus *Musa*, and facilitated the recognition of duplicates to be eliminated and clarified from uncertainties or mislabelled banana accessions introduced to the collection.

3.6. Amplified Fragment Length Polymorphism (AFLP)

AFLP is a DNA marker based on PCR amplification of selected restricted fragments obtained from the digestion of total genomic DNA or cDNA [55, 56]. It is a robust and reliable molecular technique recently employed in many systematic plant studies. AFLP banding patterns should be treated initially as dominant markers; this makes the information content limited. However, AFLP patterns can be detected as codominant markers in a segregating population when the analysis is applied to large populations [57].

The nature of an individual, whether a homozygote or heterozygote, could be distinguished using software developed on the basis of band intensity. Moreover, AFLP results in a binary band presence–absence matrix profile. In that case, two factors may affect band detection and analysis: the first is that identical bands may correspond to different fragments (homoplasy); the second is that different fragments appear as a single band (collision). An estimation method was reported for solving the effect of these factors [58], in which AFLP was demonstrated as a sampling procedure of fragments, with lengths sampled from a distribution. This study focused on estimation of pairwise genetic similarity, defined as average fraction of common fragments. Levels of polymorphism in *Musa* were shown to be high when analysed using AFLP, indicating that the technique was effective for genetic diversity analysis [59-62]. In this regard, three subspecies were suspected in the *acuminata* complex based on AFLP analysis, dominated by the subspecies *microcarpa*, *malaccensis* and *burmannica* [63].

The relationship between *M. acuminata* and *M. balbisiana* and their relatedness to cultivated bananas has been reported more clearly using AFLP [64]. Furthermore, several primers were selected from the AFLP results that can be easily used to identify A and B genomes within cultivars using a simple PCR. Additionally, AFLP markers were reported as a powerful tool for evaluating genetic polymorphisms and relationships in *Musa*. They also serve for discriminating amongst species with, A, B, S and T genomes within *Musa* species, as well as between plantains and cooking bananas [7]. Compared to other DNA markers, AFLP was shown to be a more powerful tool than RAPD for assaying genetic polymorphisms, genetic relationships and cultivar identification among the West African plantain [65]. However, SRAP markers were more informative than AFLP in giving a higher number of unique bands specific for certain genotypes [7]. AFLP markers were appropriate for demonstrating 10 markers co-segregating with the presence or absence of banana streak badnavirus infection in *Musa* hybrids [66].

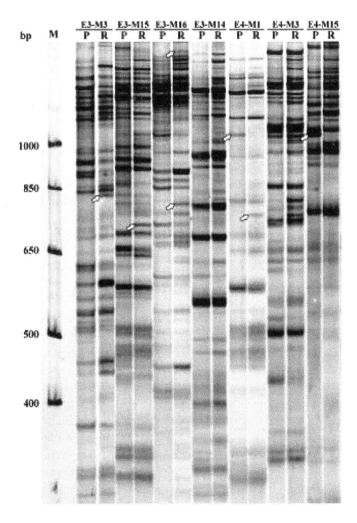

Figure 3. AFLP polymorphism of parental (P, bulk of 25 plants) and somatic embryogenesis regenerated *M. acuminata* AAA, cv. Williams triploid plants (R, bulk of 50), unique bands (arrows). The AFLP primer combinations as described by [69].

On the other hand, the AFLP marker technique was shown to be a good tool for detection of genetic variation in banana organogenesis and somatic embryogenesis-derived plants [67-69]. AFLP techniques have some disadvantages compared to other PCR-based markers, in that they are technically challenging, time consuming and relatively expensive, while requiring a number of DNA processes including digestion, ligation and amplification, as well as a complex staining system. Additionally, a relatively large amount of high-quality DNA is necessary for complete digestion, which is required to reduce the presence of fake polymorphisms. However, microsatellite markers and AFLP analysis is considered to be one of the most suitable tools for marker-assisted breeding in *Musa*. DNA markers associated with fruit parthenocarpy, dwarfism and apical dominance in banana and plantain are being identified using AFLP and SSR techniques [58, 31].

4. Single Nucleotide Polymorphism (SNP)

Single nucleotide variations in the genome sequence of individuals of a population or species are known as single nucleotide polymorphisms (SNPs). The development of this technique in humans demonstrated improvements in sequencing technology and availability of an increasing number of SNP sequences [70]; this development has made direct analysis of genetic variation at the DNA sequence level possible in genomes from different organisms [71]. Modern high-throughput DNA sequencing technologies and bioinformatics tools have led to the discovery that SNPs constitute the most abundant molecular markers in the plant genomes, which has revolutionized the pace and precision of plant genetic analysis, and the discovery that SNPs are widely distributed throughout genomes, although their occurrence and distribution varies among species [72].

4.1. Diversity Arrays Technology (DArT)

DArTs are attractive approaches to detecting large numbers of genome-specific single nucleotide polymorphism (SNP) markers [73] and EcoTILLING. In principle, DArT is a DNA hybridization-based genotyping technology, which enables low-cost whole-genome profiling of crops without prior sequence information. DArT reduces the complexity of a representative sample (such as pooled DNA representing the diversity of *Musa*) using the principle that the genomic "representation" contains two types of fragments: constant fragments, found in any "representation" prepared from a DNA sample from an individual belonging to a given cultivar or species, and variable (polymorphic) fragments called molecular markers, found in some but not all of the "representations". DArT markers are biallelic and may be dominant (present or absent) or codominant (two doses vs. one dose or absent). However, this technology has disadvantages when compared to other molecular markers, in that it depends on the availability of the array, a microarray printer and scanner, and computer infrastructure to analyse, store and manage the data produced. Despite these disadvantages, the markers are sequence-ready and, therefore, if sequenced they can be developed for a PCR analysis using standard electrophoresis.

Sequenced DArT markers have been used with *Musa* in several studies. In this regard, approximately 1,500 DArT markers have been developed using a wide array ("metagenome") of *Musa* accessions [74]. These can be attached to BAC contigs, and can simplify the construction of high-quality physical maps of the banana genome, which is a critical step in a sequencing project. On the other hand, the carotenoid content and genetic variability of some banana accessions were evaluated from the *Musa* germplasm collection held at Embrapa Cassava and Tropical Fruits, Brazil [75]. Forty-two samples were analysed, including diploids, triploids and tetraploids. The molecular analysis performed using 653 DArT markers showed that DArT was an efficient tool and revealed wide-ranging genetic variability in the collected accessions. Furthermore, DArT markers were used to analyse a panel of 168 *Musa* genotypes; thus, the genomic origin of the markers can help to resolve the pedigree of valuable genotypes of unknown origin. A total of 836 markers were identified and used for genotyping. Ten percent of them were specific to the A genome and enabled the targeting of this genome portion in relatedness analysis among diverse ploidy constitutions.

DArT revealed genetic relationships among *Musa* genotypes consistent with those provided by the other marker technologies, but at a significantly higher resolution and speed [76]. Likewise, DArTs were used for the *Musa* framework map that was developed at CIRAD; additionally, 380 of these markers have been used in the construction of the BORLI map, also at CIRAD [18]. In another study, DArT analysis was used to verify the genome constitution of 24 Philippine *Musa* cultivars. Results of the molecular data showed that some DArT markers were specific for the B genome; subsequently, these markers can identify cultivars with B genome regardless of the presence of the A genome. Hence, these markers can be used to establish genome identity of the *Musa* cultivars. Moreover, BB and BBB accessions were separated from AA/AAA and AAB cultivars in a dendrogram based on DArT data [77]. Recently, a molecular marker-based genetic linkage map of two related diploid banana populations with a complex pedigree was achieved [78], using 121 DArT markers accompanied with allele-specific-polymerase chain reactions (AS-PCR) and simple sequence repeats (SSR). The linkage analysis indicated the likely presence of structural rearrangements.

4.2. (Eco) Targeting Induced Local Lesions in Genomes (EcoTILLING)

EcoTILLING is a high-throughput method for the discovery and characterization of SNPs and small insertions/deletions (indels) in genomes [53, 79, 80]. It is an adaptation of the enzymatic mismatch cleavage and fluorescence detection methods originally developed for the targeting induced local lesions in genomes (TILLING) reverse-genetic strategy [79, 81]. The technique was first described for *Arabidopsis* ecotypes (it was therefore named EcoTILLING). It is has been used in many organisms due to its accuracy, low-cost and high-throughput routine, and it also serves for the discovery and assessment of genetic diversity. About 700–1,600 bp gene target regions are amplified using gene-specific primers that are fluorescently labelled for EcoTILLING using enzymatic mismatch cleavage. After PCR, samples are denatured and annealed, and heteroduplexed molecules are created through the hybridization of polymorphic amplicons. Mismatched regions in otherwise double-stranded duplex are then cleaved using a crude extract of celery juice containing the single-strand specific nuclease CEL I. Cleaved products are resolved by denaturing polyacrylamide gel electrophoresis (PAGE) and observed by fluorescence detection.

The EcoTILLING method was used for the discovery and characterization of nucleotide polymorphisms in *Musa*, including diploid and polyploid accessions. Over 800 novel alleles were discovered in 80 accessions, indicating that EcoTILLING is a robust and accurate platform for the discovery of polymorphisms in homologous gene targets. The method proved to be valid in identifying two SNPs that might be deleterious for the function of an important gene in phototropism. Using a principle component analysis, it was shown that evaluation of heterozygous SNPs alone was sufficient to discriminate hybrids from non-hybrids and triploid *acuminata* plants from diploids. Thus, a rapid SNP assessment can in some cases replace flow cytometric methods used to differentiate ploidy.

Furthermore, differentiation between *acuminata* and *balbisiana* diploids was adequate to uncover an accidental miss-assignment of an AA type as a BB type by the stock centre. Moreover, a high level of nucleotide diversity in *Musa* accessions was revealed by evaluating the heterozygous polymorphism and haplotype blocks. The authors concluded that EcoTIL-LING was an accurate and efficient method for the detection and classification of nucleotide polymorphisms in diploid and polyploid banana. It is highly scalable and many applications can be considered, from simple measurements of heterozygosity as a selection criterion in breeding programmes to more nuanced studies of chromosomal inheritance and functional genomics analysis. This strategy can be used to develop hypotheses for inheritance patterns of nucleotide polymorphisms within and between genome types [53].

More recently, single nucleotide polymorphism (SNP) studies for marker discovery of the use of beta carotene (provitamin A) in plantains [82], and SNPs found in the partial sequence of the gene encoding the large sub-units of ADP-glucose pyrophosphorylase, a key enzyme related to starch metabolism, in banana and plantains [83], give important information for new approaches to investigating the wide range of banana germplasm biodiversity and incorporating the information in banana and plantain breeding.

5. Molecular cytogenetic

Using cytogenetics, chromosome studies of humans and several plant species have been going on for more than a century, helping to establish the typical number of chromosomes for each of the species and to assign a chromosome number according to their size and centromere position. Chromosome size and banding pattern helped to identify subchromosomal regions and were associated with some phenotypic characteristics. Chromosome number has also been important for identifying individuals among species. Even though cytogenetic tools have been improved to obtain high-resolution banding patterns for identifying deletions, insertions or translocations [84], it remains a challenge to elucidate the origins of the chromosomes that are involved in chromosome rearrangements. Molecular cytogenetics is adding a set of powerful tools to those already available for studying genome organization, evolution and recombination. This technology can help to identify small changes at the level of the gene, for which several techniques have been developed.

5.1. Fluorescent *in situ* Hybridization (FISH) to chromosomes

Fluorescent *in situ* hybridization (FISH) allows hybridization sites to be visualized directly and, moreover, several probes can be simultaneously detected with different fluorochromes, allowing the physical order of the chromosomes to be determined. FISH was used on mitotic chromosomes to localize the physical sites of 18S-5.8S-25S and 5S rRNA genes in *Musa* [85, 86]. A single major intercalary site was observed on the short arm of the nucleolar organizing chromosome in both A and B genomes. Diploid, triploid and tetraploid genotypes showed two, three and four sites, respectively. Heterogeneous *Musa* lines showed different intensity of signals that indicate variation in the number of copies of these genes. In the case of 5S rDNA, eight subterminal sites were observed in Calcutta 4 (AA), while Butohan 2 (BB) had six sites. Triploid lines showed six to nine major sites of 5S rDNA of widely varying intensity, near the limit of detection. The diploid hybrids had five to nine sites of 5S rDNA while the tetraploid hybrid had 11 sites [86].

Additionally, a dual colour FISH showed that in all studied accessions, the satellite chromosomes carrying the 18S-25S loci did not carry the 5S loci [85]. On the other hand, the telomeric sequence was detected as pairs of dots at the ends of all the chromosomes analysed, but no intercalary sequences were seen [85].

Detection of the integration of viral sequences of banana streak badnavirus (BSV) in two metaphase spreads of *Obino l'Ewai* plantain (AAB) was achieved using FISH [87]. Two different BSV sequence locations were revealed in *Obino l'Ewai* chromosomes and a complex arrangement of BSV and *Musa* sequences was shown by probing stretched DNA fibres. In another study, the *monkey* retrotransposon was identified and localized in *Musa* using FISH [88]. Several copies of *monkey* were concentrated in the nucleolar organizer regions and colocalized with rRNA genes. Other copies of *monkey* appear to be dispersed throughout the genome. In addition, in order to increase the number of useful cytogenetic markers for *Musa*, low amounts of repetitive DNA sequences of BAC clones were used as probes for FISH on mitotic metaphase chromosomes [89]. Only one clone gave a single-locus signal on chromosomes of *M. acuminata* cv. Calcutta 4. The clone localized on a chromosome pair that carries a cluster of 5S rRNA genes. The remaining BAC clones gave dispersed FISH signals throughout the genome and/or failed to produce any signal. In addition, 19 BAC clones were subcloned and their 'low-copy' subclones were selected, to avoid the excessive hybridization of repetitive DNA sequences. Out of these, one subclone gave a specific signal in secondary constriction on one chromosome pair, and three subclones were localized into centromeric and peri-centromeric regions of all chromosomes. The nucleotide sequence analysis revealed that subclones which localized on different regions of all chromosomes contained short fragments of various repetitive DNA sequences [89].

Furthermore, a modern chromosome map technology known as high-resolution fluorescent *in situ* hybridization (FISH) was applied in *Musa* species using BAC clone positioning on pachytene chromosomes of Calcutta 4 (*M. acuminata, Eumusa*) and *M. velutina (Rodochlamys)*. To make cell spread preparations appropriate for FISH, pollen mother cells were digested with pectolytic enzymes and macerated with acetic acid. BAC clones that contain markers for known resistance genes were chosen and hybridized to establish their relative positions on the two species [90]. Centromeric retrotransposons were detected in banana chromosomes hybridized with MusA1 by FISH. Since all of the banana chromosomes are metacentric or submetacentric, signals were located around the centre of the chromosome, indicating that loci are found near or within the centromere [91]. Recently, the genomic organization and molecular diversity of two main banana DNA satellites were analysed in a set of 19 *Musa* accessions, including representatives of A, B and S genomes and their interspecific hybrids. The two DNA satellites showed a high level of sequence conservation within, and a high homology between *Musa* species. FISH with probes for the satellite DNA sequences, rRNA genes and a single-copy BAC clone 2G17 resulted in characteristic chromosome banding patterns in *M. acuminata* and *M. balbisiana*, which may aid in determining genomic constitution in interspecific hybrids. In addition, the knowledge of *Musa* satellite DNA was improved by increasing the number of cytogenetic markers and the number of individual chromosomes which can be identified in *Musa* [92].

5.2. Genomic *in situ* Hybridization (GISH)

Genomic *in situ* hybridization (GISH) is a powerful tool to differentiate alien chromosomes or chromosome segments from parental species in interspecific hybrids [93]. It has been applied to mitotic chromosomes from many plants resulting from interspecific hybridization. However, application of GISH on meiotic chromosomes is challenging, and has been reported in just a few species. In *Musa*, GISH was successfully applied to differentiate the chromosomes of different genomes on both mitotic and meiotic chromosomes.

The first study that used GISH in *Musa* was with chromosome spreads prepared from root tips. The total genomic DNA of diploid lines AA and BB was able to label the centromeric regions of all 22 chromosomes of the corresponding line. However, the two satellite chromosomes of genome B labelled strongly with genomic A DNA. GISH discriminated between A and B chromosomes in AAB and ABB cultivars. Additionally, it has immense potential for identification of chromosome origin and can be used to characterize cultivars and hybrids produced in *Musa* breeding [94]. In another study [95], GISH was used to determine the exact genome structure of interspecific cultivated clones AAB and ABB. As a notable exception, the clone 'Pelipita' (ABB) was found to have eight A and 25 B chromosomes instead of the predicted 11 A and 22 B. In addition, chromosome complement was determined by some clones that could not be classified by phenotypic characteristics and chromosome counts. Moreover, rDNA sites were located in *Musa* species that appeared to be frequently associated with satellites. These sites can be separated from the chromosomes providing a potential source of chromosome counting errors when conventional techniques are used. Furthermore, GISH successfully differentiated the chromosomes of the four known genomes, A, B, S and T, which correspond to the genetic constitutions of wild *Eumusa* species *M. acuminata*, *M. balbisiana*, *M. schizocarpa* and the *Australimusa* species, respectively [95, 96].

On the other hand, GISH has been used on meiotic chromosomes in plants. It is quite challenging, and the protocols used are complex and highly variable depending on the species. A method was developed to prepare chromosomes at meiosis metaphase I suitable for GISH in *Musa* [97]. The main challenge encountered was the hardness of the cell wall and the density of the microsporocyte's cytoplasm, which hamper the accessibility of the probes to the chromosomes and generate higher levels of background noise. It was clearly demonstrated that interspecific recombinations between *M. acuminata* and *M. balbisiana* chromosomes do occur and may be frequent in triploid hybrids. Recently, conventional cytogenetic and GISH analyses of meiotic chromosomes were used to investigate the pairing of different chromosome sets at diploid and tetraploid levels, and to reveal the chromosome constitution of hybrids derived from crosses involving allotetraploid genotype. At both ploidy levels, the analysis suggested that the newly formed allotetraploid behaves as a segmental allotetraploid. The 11 chromosomes were found as three sets in a tetrasomic pattern, three in a likely disomic pattern and the five remaining sets in an intermediate pattern. In addition, balanced and unbalanced diploid gametes were detected in progenies. The chromosome constitution was more homo-genous in pollen than in ovules. The segmental inheritance pattern exhibited by the AABB allotetraploid genotype implies chromosome exchanges between *M. acuminata* and *M. balbisiana* species, and opens new horizons for reciprocal transfer of valuable alleles [97].

5.3. Flow cytometry

Flow cytometry (FCM) protocols have been applied for studying the natural variation in *Musa* nuclear genome size (DNA content) for taxonomic purposes and for checking ploidy among gene bank accessions and breeding materials [98-101]. The literature data suggest that, on average, the A genome of *M. acuminata* and clones with AA genome constitution is around 12 % larger than the B genome of *M. balbisiana*, with small intraspecific variation in nuclear DNA found in a number of wild acuminata diploid and parthenocarpic bananas, whereas large variation seems to be exhibited among triploid varieties [102, 99].

The study of genomic composition of *Musa* accessions on a core collection based on ITS (internal transcribed spacer sequences of the nuclear ribosomal DNA) regions and SSR polymorphism, along with assessment of DNA content and ploidy by FCM, has given support to the hypothesis [103] of the occurrence of homologous recombination between A and B genomes, or between *M. acuminata* subspecies genomes, leading to discrepancies in the number of sets or portions from each parental genome [104]. It is worth mentioning that when the former published data sets were compared, clear differences in the genome size of *M. balbisi-ana* accessions were found, leaving open the question as to the origin of such variation. Basic research works of this type are of great interest to understanding the evolution and domesti-cation of *Musa*, and for betterment of bananas using new approaches.

Research on the genetic stability/instability of *in vitro* somatic embryogenesis (SE) cultures by FCM for polyploid bananas [105, 106], and recently, by FCM and cytological analysis of embryogenic *M. acuminata* ssp. *malaccensis* cell suspension cultures, and of their somatic embryo-derived plantlets [107], adds support to the use of *in vitro* innovative cell biology tools for high-throughput production of clean banana planting materials, the rescue of hybrid true seeds by *in vitro* germination and their clonal propagation through SE, which assist in the acquisition of seedlings after interspecific hybridization and across ploidy to enhance the efficiency of genetic improvement in *Musa*.

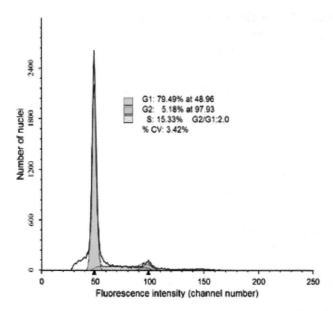

Figure 4. Typical histogram of propidium iodide (PI)-labelled nuclei at pre-DNA synthesis (G1), synthesis (S) and post-synthesis (G2) from a triploid (2n=3x=33 chromosome) regenerated *M. acuminata* AAA, cv. Williams plant. The x axis indicates the florescence signal intensity (canal number), which stoichiometrically relates to DNA content. Plant material was regenerated by somatic embryogenesis [68].

Criterion	RFLP	RAPD	VNTR	ISSR	SRAP	AFLP	SNPs
Quantity of DNA required	High	Low	Low/high[a]	Low	Low	Moderate	Low
Quantity of information	Low	High	High	High	High	High	Low
Replicability	High	Variable	High	High	High	High	Moderate
Resolution of genetic differences	High	Moderate	High	Moderate	High	High	High
Abundance	High	High	Medium	High	High	High	High
Locus specificity	Yes	No	Yes	No	Yes	No	Yes
Ease of use and development	Difficult	Easy	Easy	Easy	Easy	Moderate[b]	Moderate
Development time	Long	Short	Moderate	Short	Short	Moderate	Moderate
Development costs	High	Low	Moderate	Low	Low	Moderate	Low
Operational costs	High	Low	Low	Low	Low	Moderate	Moderate / High
Locus specificity	Yes	No	Yes	No	Yes/No	No	Yes
Major application	Physical mapping	Gene tagging	Genetic diversity	Gene tagging	Genetic diversity	Gene tagging	SNP mapping

[a]VNTR requires a high quantity of DNA when analysed by RLFP, while it needs a low quantity when analysed by PCR. [b]Analysis of AFLP is easy with the help of an automated genotyper, or when using low-resolution agarose gel electrophoresis, but manual polyacrylamide electrophoresis requires a certain amount of experience

Table 1. Comparison among different molecular markers for various criteria; adapted from [22, 51, 108].

6. Protein polymorphism

Proteomics can be defined as the systematic analysis of proteome, the protein complement of the genome, which deals with information on proteins' abundance, their variations and modifications, and their interacting partnerships and networks, in order to understand cellular processes in biological systems. Thus, proteomics is important for understanding the molecular mechanisms involved in plant and crop biodiversity, which is a driving force behind speciation, crop domestication and improvement. This is of particular interest in bananas, which are good representatives of a complex allopolyploid and an important fruit crop.

6.1. Proteome analysis

In recent years, several proteomic studies, based on combined use of two-dimensional electrophoresis (2DE) and mass spectrometric methods, have been successfully applied to investigate the effect of osmotic stresses on banana growth and development [109], cold tolerance [110], inter-and intra-cultivar protein polymorphisms [111], the fruit proteome of banana [112, 113], and the proteomic profiling of banana roots in response to *F. oxysporum* [114]. These investigations highlight the value of new sequencing technologies for integrating the biological information of a plant system usually considered a non-model crop, and open new approaches to studying biodiversity, evolution and domestication in *Musa*. In this scenario, the proteomic analysis of shoot meristem changes in the acclimation to sucrose-mediated osmotic stress of two banana varieties uncovered several genotype-specific proteins (isoforms), enzymes of the energy metabolism (e.g., phosphoglyceate kinase, phosphogluco-mutase, UDP-glucose pyrophosphorylase) and stress adaptations (e.g., OSR40-like protein, abscisic stress ripening protein-like protein, ASR) that were associated with the dehydration-tolerant variety [109].

In contrast, comparative quantitative proteomic analysis of plantain (AAB genome) response to cold stress treatments revealed that about 23.3 % of the 3477 total proteins identified were differentially expressed. The largest parts of the expressed proteins were predicted to be involved in the oxidation reduction process (including oxylipin biosynthesis), cellular process, response to stress and primary metabolic process. Interestingly, among the cold-responsive proteins involved in the oxidation reduction process, Cu/Zn SOD (superoxide dismutase), CAT 2 (catalase isozyme 2) and LOX (lipoxygenease) were found to be differentially expressed in the cold-tolerant plantain, in contrast to the cold-susceptible banana [110]. Altogether, the previous works provided clues as to the existence of inter-variety protein polymorphism related to their *acuminata* or *balbisiana* origin, and open new approaches to examining diploid and triploid bananas' different responses to environmental stress.

Moreover, evidence from the proteome analysis of different triploid banana varieties using 2D electrophoresis revealed the following results: i) principal component analysis (PCA) showed that the principal component PC1 (which explains 39 % of the variance information) was positively correlated with the presence of the B genome; and, ii) the hierarchical clustering revealed that the first level of clustering separates the varieties of genome BBB and ABB composition from both AAB and the two AAA varieties, the second level splits both AAB varieties from the two AAA varieties and the ABB from the BBB varieties, and the third level divides both AAA varieties and both AAB varieties. Although proteome analysis does not always correspond to the presumed genome formulae, perhaps because following polyploidization new gene copies may undergo modifications allowing functional diversification, in general, the observations at the protein level provide good indications for a more complex genome structure and genomic rearrangement in some banana varieties [111].

7. Conclusion

The limited genetic knowledge of the banana genome and the nature of the crop as a parthenocarpic fruit and a mostly triploid, sterile plant mean that many aspects of breeding and selection that have been possible in other crops cannot be applied in the banana. Several approaches to breeding and selection have been applied in numerous plant species; however, they could not be used in banana due to the unclear genetic knowledge of its genome and the natural characteristics of the crop, including parthenocarpy, ploidy and sterility. Unconventional biotechnological strategies including DNA and protein-based marker techniques have contributed considerably in providing a vast amount of information that helps in understanding the nature of *Musa* genome and its genetic diversity. DNA-based markers were developed and are being used in *Musa*, representing powerful tools to assess the genetic diversity and clarify the individual genetic characteristics and relationships. Researchers should select the best marker for a certain task; however, the recently developed molecular markers were more informative when applied in banana. For instance, SRAPs showed better assessment of the genetic diversity and increased the clarity of genetic relationships among several *Musa* species, subspecies and cultivars when compared with other traditional markers, such as RAPD and AFLP. Moreover, since SRAP is based on the amplification of ORFs, this gives another advantage to this marker over others. Involving recently developed molecular markers, such as the target region amplification polymorphism (TRAP), which is based on the amplification of target EST sequences, as well as the intron targeted amplified polymorphism (ITAP), which is based on the amplification of 3′ widely distributed intron-exon splice junction sequences, could improve the assessment of genetic diversity in *Musa* and refine the genetic relationships among the different sections, species and mislabelled accessions. In addition, despite their relatively high cost, the high-throughput technologies based on SNPs or small-scale indels are efficient alternatives to traditional markers, because of their greater abundance, high polymorphism, ease of measurement and ability to reveal hidden polymorphisms where other methods fail. SNPs also allow easy and unambiguous identification of alleles or haplotypes. A good marker system for polyploid crops should be dosage sensitive and have the ability to distinguish heterozygous genotypes with multiple haplotypes. On the other hand, molecular cytogenetics techniques have played a great role in understanding the *Musa* genome construction, determining genomic constitution of the interspecific hybrids and studying the natural variation in *Musa* nuclear genome size, as well as checking ploidy levels among gene bank accessions and breeding materials. In addition, the knowledge on *Musa* satellite DNA was improved by increasing the number of cytogenetic markers and the number of individual chromosomes which can be identified in *Musa*. The complementation of DNA based markers with protein-based ones has completed the image of unreachable variations at DNA level. Several proteomic studies, based on combined use of two-dimensional electrophoresis (2DE) and mass spectrometric methods, have been successfully applied to investigate various approaches in banana, including protein polymorphism, and biotic and abiotic tolerance. The combination of all molecular approaches surveyed and discussed in this chapter can help in the revelation of the genetic diversity in *Musa*. High-throughput technologies based on SNPs or small-scale indels are efficient alternatives for traditional markers (RFLP, RAPD or AFLP), because of their greater abundance, high polymorphism, ease of measurement and ability to reveal hidden polymorphisms where other methods fail. SNPs also allow easy and unambiguous identification of alleles or haplotypes. A good marker system for polyploid crops should be dosage sensitive, and have the ability to distinguish heterozygous genotypes with multiple haplotypes.

Author details

G. Manzo-Sánchez[1*], M.T. Buenrostro-Nava[1], S. Guzmán-González[1], M. Orozco-Santos[2], Muhammad Youssef[3] and Rosa Maria Escobedo-Gracia Medrano[4]

*Address all correspondence to: gmanzo@ucol.mx

1 School of Agronomy and Biological Sciences, University of Colima; Tecoman, Colima, Mexico

2 INIFAP-Tecoman Experimental Station, Colima, Tecoman, Colima, Mexico

3 Department of Genetics, Faculty of Agriculture, Assiut University, Assiut, Egypt

4 Plant Biochemistry and Molecular Biology Unit, Yucatan Center for Scientific Research, Chuburna de Hidalgo, Merida, Yucatan, Mexico

References

[1] International Institute of Tropical Agriculture. Banana & Plantain. http://www.iita.org/banana-and-plantain (accessed September 7th, 2014)

[2] Ortiz R, Swennen R. From cross breeding to biotechnology-facilitated improvement of banana and plantain. Biotechnology Advances 2014; 32(1) 158-169.

[3] Feuillet C, Leach JE, Rogers J, Schnable PS, Eversole K. Crop genome sequencing: Lessons and rationales. Trends in Plant Sciences 2011; 16(2) 77-88.

[4] Liu BH. 1997. Statistical Genomics: Linkage, Mapping and QTL Analysis. CRC Press, 1997, p. 611.

[5] D'Hont A, Denoeud F, Aury JM, Baurens FC, Carreel F, Garsmeur O, Noel B, Bocs S, Droc G, Rouard M, Da Silva C, Jabbari K, Cardi C, Poulain J, Souquet M, Labadie K, Jourda C, Lengellé J, Rodier-Goud M, Alberti A, Bernard M, Correa M, Ayyampa-layam S, Mckain MR, Leebens-Mack J, Burgess D, Freeling M, Mbéguié-A-Mbéguié D, Chabannes M, Wicker T, Panaud O, Barbosa J, Hribova E, Heslop-Harrison P, Ha-bas R, Rivallan R, Francois P, Poiron C, Kilian A, Burthia D, Jenny C, Bakry F, Brown S, Guignon V, Kema G, Dita M, Waalwijk C, Joseph S, Dievart A, Jaillon O, Leclercq J, Argout X, Lyons E, Almeida A, Jeridi M, Dolezel J, Roux N, Risterucci AM, Weis-senbach J, Ruiz M, Glaszmann JC, Quétier F, Yahiaoui N, Wincker P. The banana (*Musa acuminata*) genome and the evolution of monocotyledonous plants. Nature 2012; 488 213-217.

[6] Davey MW, Gudimella R, Harikrishna JA, Sin LW, Khalid N, Keulemans J. A draft *Musa balbisiana* genome sequence for molecular genetics in polyploid, inter-and intra-specific *Musa* hybrids. BMC Genomics 2013; 14 683. http://www.biomedcentral.com/1471-2164/14/683 (accesed 20 October 2014).

[7] Youssef M, James A, Rivera-Madrid R, Ortiz R, Escobedo-Gracia Medrano RM. *Musa* genetic diversity revealed by SRAP and AFLP. Mol Biotech 2011; 47(3) 189–199.

[8] Valdez-Ojeda R, James-Kay A, Ku-Cauich J, Escobedo-Gracia Medrano RM. Genetic relationships among a collection of *Musa* germplasm by fluorescent-labeled SRAP. Tree Genetics and Genomes 2014; 10 (3) 465-476.

[9] Heslop-Harrison JS, Schwarzacher T. Domestication, genomics and the future for ba-nana. Annals of Botany 2007; 100 (5) 1073-1084.

[10] Botstein D, White RL, Skolnick M, Davis RW. Construction of a genetic linkage map in man using restriction fragment length polymorphisms. Am J Hum Genet 1980; 32 (3) 314-333.

[11] Bhat KV, Jarret RL. Random amplified polymorphic DNA and genetic diversity in Indian *Musa* germplasm. Genet Resour Crop Evol 1995; 42 (2) 107–118.

[12] Gawel NJ, Jarret RL, Whittemore AP. Restriction fragment length polymorphism (RFLP)-based phylogenetic analysis of *Musa*. Theor Appl Genet 1992; 84 (3-4) 286–290.

[13] Jarret R, Gawel N, Whittemore A, Sharrock S. RFLP-based phylogeny of *Musa* spe-cies in Papua New Guinea. Theor Appl Genet 1992; 84 (5-6) 579–584.

[14] Lanaud C, Tezena du Montcel H, Jolivot MP, Glaszman JC, Gonzalez de León D. Variation in ribosomal gene spacer length among wild and cultivated banana. He-redity 1992; 68 148-156.

[15] Jenny C, Carreel F, Bakry F. Revision of banana taxonomy: Klue Tiparot (*Musa* sp.) reclassified as a triploid. Fruits 1997; 52 (2) 83-91.

[16] Gawel NJ, Jarret RL. Chloroplast DNA restriction fragments length polymorphisms (RFLPs) in *Musa* species. Theor Appl Genet 1991; 81 (6) 783–786.

[17] Baurens FC, Noyer JL, Lanaud C, Lagoda PJL. Sequence tagged site markets to draft the genomic structure of the banana chloroplast. Fruits 1997; 52 (4) 247-259.

[18] Hippolyte I, Bakry F, Seguin M, Gardes L, Rivallan R, Risterucci A-M, Jenny C, Perrier X, Carreel F, Argout X, Piffanelli P, Khan I, Miller R, Pappas G, Mbeguie-A-Mbeguie D, Matsumoto T, De Bernardinis V, Huttner E, Kilian A, Baurens F-C, D'Hont A, Cote F, Courtois B, Glaszmann JC. A saturated SSR/DArT linkage map of *Musa acuminata* addressing genome rearrangements among bananas. BMC Plant Biol 2010; 10: 65–65. http://www.biomedcentral.com/1471-2229/10/65 (accesed 20 October 2014).

[19] Uma S, Siva SA, Saraswathi MS, Manickavasagam M, Durai P, Selvarajan R Sathiamoorthy S. Variation and intraspecific relationships in Indian wild *Musa balbisiana* (BB) population as evidenced by random amplified polymorphic DNA. Genetic Resources in Crop Evolution 2006; 53 (2) 349-355.

[20] Pillay M., Ashokkumar K., James A., Prince Kirubakara SJ, Miller R, Ortiz R, Sivalingam E. Molecular markers techniques in *Musa* genomic research. In: Genetics, Genomics, and Breeding of Bananas, ed. Pillay M, Ude G and Kole C: Science Publishers; 2012. pp. 70-90.

[21] Williams JGK, Kubelik AR, Livak KJ, Rafalski JA, Tingey SV. DNA polymorphisms amplified by arbitrary primers are useful as genetic markers. Nucleic Acids Res 1990; 18 (22) 6531–6535.

[22] Agarwal M, Shrivastava N, Padh H. Advances in molecular marker techniques and their applications in plant sciences. Plant Cell Reports 2008; 27 (4) 617-631.

[23] Kaemmer D. Oligonucleotide and amplification fingerprinting of wild species and cultivars of banana (*Musa* spp.) Bio/Technology 1992; 10 (9) 1030-1035.

[24] Howell EC, Newbury HJ, Swennen RL, Withers LA, Ford-Lloyd BV. The use of RAPD for identifying and classifying *Musa* germplasm. Genome 1994; 37 (2) 328–332.

[25] Bhat KV, Jarret RL. Random amplified polymorphic DNA and genetic diversity in Indian *Musa* germplasm. Genet Resour Crop Evol 1995; 42 (2)107–118.

[26] Jain PK, Saini MK, Pathak H, Gupta VK. Analysis of genetic variation in different banana (*Musa* species) variety using random amplified polymorphic DNAs (RAPDs). Afri J Biotechnol 2007; 6(17) 1987–1989.

[27] Ray T, Dutta I, Saha P, Das S, Roy SC. Genetic stability of three economically important micropropagated banana (*Musa* spp.) cultivars of lower Indo-Gangetic plains, as assessed by RAPD and ISSR markers. Plant Cell Tiss Org Cult 2006; 85 (1) 11–21.

[28] Bairu MW, Fennell CW, van Staden J. The effect of plant growth regulators on soma-clonal variation in Cavendish banana (*Musa* AAA cv. 'Zelig'). Sci Hort 2006; 108 (4) 347–351.

[29] Lakshmanan V, Venkataramareddy SR, Neelwarne B. Molecular analysis of genetic stability in long-term micropropagated shoots of banana using RAPD and ISSR markers. Elec J Biotechnol 2007; 10 (1) 106–113.

[30] Imelda M, Estiati A, Hartati NS. Induction of mutation through gamma radiation in three cultivars of banana. Ann Bogorien 2001; 7 (2) 75–82.

[31] Hautea DM, Molina GC, Balatero CH, Coronado NB, Perez EB, Alvarez MTH, Cana-ma AO, Akuba RH, Quilloy RB, Frankie RB, Caspillo CS. Analysis of induced mu-tants of Philippine with molecular markers. In: Banana Improvement: Cellular, Molecular Biology and Induced Mutations, ed. Jain S, Swennen R. Kluwer Academic Publishers, Dordrecht, the Netherlands; 2004. p 45–57.

[32] Faure S, Noyer JL, Horry JP, Bakry F, Lanaud C, Goazalez de Lean D. A molecular marker-based linkage map of diploid bananas (*Musa acuminata*). Theor Appl Genet 1993; 87(4) 517-526.

[33] Pillay M, Nwakanma DC, Tenkouano A. Identification of RAPD markers linked to A and B genome sequences in *Musa* L. Genome 2000; 43 (3) 763–767.

[34] Pillay M, Ogundiwin E, Tenkouano A, Dolezel J. Ploidy and genome composition of *Musa* germplasm at the International Institute of Tropical Agriculture (IITA). Afr J Biotechnol 2006; 5(13) 1224–1232.

[35] Crouch HK, Crouch JH, Jarret RL, Cregan PB, Ortiz R. Segregation at microsatellite loci in haploid and diploid gametes of *Musa*. Crop Science 1998; 38(1) 211–217.

[36] Das BK, Jena RC, Samal KC. Optimization of DNA isolation and PCR protocol for RAPD analysis of banana/plantain (*Musa* spp.). Int J Agric Sci 2009; 1(2) 21–25.

[37] Khatri A, Dahot MU, Khan I, Raza S, Bibi S, Yasmin S, Nizamani DGS. *In vitro* muta-genesis in banana and variant screening through ISSR. Pak J Bot 2011; 43(5) 2427-2431.

[38] Nadal-Medina R, Manzo-Sanchez G, Orozco-Romero J, Orozco-Santos M, Guzman-Gonzalez S. Genetic diversity of bananas and plantains (*Musa* spp.) determined by RAPD markers. Revista Fitotecnia Mexicana 2009; 32(1) 1–7.

[39] Tourmente S, Deragon JM, Lafleuriel J, Tutois S, Pelissier T, Cuvillier C, Espagnol MC, Picard G. Characterization of minisatellites in *Arabidopsis thaliana* with sequence similarity to the human minisatellite core sequence. Nucl Acid Res 1994; 22(16) 3317–3321.

[40] Kaemmer D, Fischer D, Jarret RL, Baurens FC, Grapin A, Dambier D, Noyer JL, La-naud C, Kahl G, Lagoda PJL. Molecular breeding in the genus *Musa*: A strong case for STMS marker technology. Euphytica 1997; 96(1) 49–63.

[41] Cheung F, Town CD. A BAC end view of the *Musa* acuminata genome. BMC Plant Biol 2007; 7 29–29. http://www.biomedcentral.com/1471-2229/7/29 (accesed 20 October 2014).

[42] Jarret RL, Bhat KV, Cregan PB, Ortiz R, Vuylsteke D. Isolation of microsatellite DNA markers in *Musa*. InfoMusa 1994; 3(2) 3–4.

[43] Wang JY, Zheng LS, Huang BZ, Liu WL, Wu YT. Development, characterization, and variability analysis of microsatellites from a commercial cultivar of *Musa acuminata*. Genet Resour Crop Evol 2010; 57 (4) 553–563.

[44] Miller RNG, Passos MAN, Menesez NNP, Souza MT Jr, do Carmo Costa MM, Rennó Azevedo VCR, Amorim EP, Pappas GJ Jr, Ciampi AY. Characterization of novel microsatellite markers in *Musa acuminata* subsp. *burmannicoides*, var Calcutta 4. BMC Res Notes 2010; 3: 148. http://www.biomedcentral.com/1756-0500/3/148 (accesed 20 October 2014).

[45] Garcia SAL, Talebi R, Ferreira CF, Vroh BI, Paiva LV, Kema GHJ, Souza MT Jr. Identification and validation of est-derived molecular markers, TRAP and VNTRS, for banana research. Acta Hort 2011; 897: 69-80. http://www.actahort.org/books/897/897_6.htm (accesed 20 October 2014).

[46] Zietkiewicz E, Rafalski JA, Labuda D. Genome fingerprinting by simple sequence repeat (SSR)-anchored polymerase chain reaction amplification. Genomics 1994; 20(2) 176–183.

[47] Rout GR, Senapati SK, Aparajita S, Palai SK. Studies on genetic identification and genetic fidelity of cultivars banana using ISSR marker. Plant Omics Journal 2009; 2(6): 250-258.

[48] Qin XQ, Peng HX, Long X, Yao JY. Preliminary study on ISSR analysis and classification of wild *Musa* germplasm in Guangxi, China. Acta Hort 2011; 897: 259-262. http://www.actahort.org/books/897/897_29.htm (accesed 20 October 2014).

[49] Lu Y, Xu WH, Xie YX, Zhang Z, Pu JJ, Qj YX, Li HP. Isolation and characterization of nucleotide-binding site and C-terminal leucine-rich repeat-resistance gene candidates in bananas. Genet Mol Res 2011; 10(4) 3098-3108.

[50] Padmesh P. Mukunthakumar S. Vineesh PS, Hari Kumar K, Krishnan PN. Exploring wild genetic resources of *Musa acuminata* Colla distributed in the humid forests of southern Western Ghats of peninsular India using ISSR markers. Plant Cell Reports 2012; 31(9) 1591-1601.

[51] Li G, Quiros CF. Sequence-related amplified polymorphism (SRAP), a new marker system based on a simple PCR reaction: Its application to mapping and gene tagging in Brassica. Theor Appl Genet 200; 103(2-3) 455–461.

[52] Phothipan S, Silayoi B, Wanichkul K, Apisitwanich S. Genetic relationship among banana in AA, AAB and B groups using random amplified polymorphic DNA (RAPD) and sequence related amplified polymorphism (SRAP) techniques. Kasetsart J Nat Sci 2005; 39 703–710. http://kasetsartjournal.ku.ac.th/kuj_files/2008/A0804281451054134.pdf (accesed 20 October 2014).

[53] Xie Z-S, Liu D-B, Wei S-X, Chen Y-Y, Xie J-J, Wei J-Y. Analysis of genetic diversity of 29 banana genotypes by SRAP.Acta Botanica Boreali-Occidentalia Sinica 2012; 8 009. http://caod.oriprobe.com/articles/31027308/Analysis_of_Genetic_Diversity_of_29_Banana_Genotypes_by_SRAP.htm (accesed 20 October 2014).

[54] Till BJ, Jankowicz-Cieslak J, Sagi L, Huynh OA, Utsushi H, Swennen R, Terauchi R, Mba C. Discovery of nucleotide polymorphisms in the *Musa* gene pool by Ecotilling. Theor Appl Genet 2010; 121(7) 1381–1389.

[55] Vos P, Hogers R, Bleeker M, Reijans M, van de Lee T, Hornes M, Frijters A, Pot J, Peleman J, Kuiper J, Zabeau M. AFLP: A new technique for DNA fingerprinting. Nucleic Acids Res 1995; 23(21) 4407–4414.

[56] Bachem CWB, van der Hoeven RS, de Bruijn SM, Vreugdenhil D, Zabeau M, Visser RGF. Visualization of differential gene expression using a novel method of RNA fingerprinting based on AFLP: Analysis of gene expression during potato tuber development. The Plant Journal 1995; 9(5) 745-753.

[57] Masiga DK, Turner CMR. 2004. Amplified (restriction) fragment length polymorphism (AFLP) analysis. In: Methods in Molecular Biology, ed., Melville SE. Humana Press, USA, 2004, p173–186.

[58] Gort G, van Hintum V, van Eeuwijk F. Homoplasy corrected estimation of genetic similarity from AFLP bands, and the effect of the number of bands on the precision of estimation. Theor Appl Genet 2009; 119(3) 397–416.

[59] Crouch JH, Crouch HK, Constand H, Van Gysel A, Breyne P, Van Montagu M, Jarret RL, Ortiz R. Comparison of PCR-based molecular marker analysis of *Musa* breeding populations. Mol Breed 1999; 5(3) 233–244.

[60] Loh JP, Kiew R, Set O, Gan LH, Gan YY. Amplified fragment length polymorphism fingerprinting of 16 banana cultivars (*Musa* cvs.). Mol Phylogenet Evol 2000; 17(3) 360–366.

[61] Wong C, Kiew R, Loh JP, Gan LH, Lee SK, Ohn S, Gan YY. Genetic diversity of the wild banana *Musa acuminata* Colla in Malaysia as evidenced by AFLP. Ann Bot 2001; 88(6) 1017–1025.

[62] Wong C, Kiew R, Argent G, Ohn S, Lee SK, Gan YY. Assessment of the validity of the sections in *Musa* (*Musaceae*) using AFLP. Ann Bot 2002; 90(2) 231-238.

[63] U de G, Pillay M, Nwakanma A, Tenkouano A. Analysis of genetic diversity and selectional relationships in *Musa* using AFLP markers. Theor Appl Genet 2002; 104(8) 1239–1245.

[64] Wongniam S, Somana J, Swangpol S, Seelanan T, Chareonsap P, Chadchawan S, Jenjittikul T. Genetic diversity and species-specific PCR-based markers from AFLP analyses of Thai bananas. Biochemical Systematics and Ecology 2010; 38(3) 416–427.

[65] Ude G, Pillay M, Ogundiwin E, Tenkouano A. Genetic diversity in an African plantain core collection using AFLP and RAPD markers. Theor Appl Genet 2003; 107(2) 248–255.

[66] Lheureux F, Carreel F, Jenny C, Lockhart BEL, Iskra C. 2003. Identification of genetic markers linked to banana streak disease expression in inter-specific Musa hybrids. Theor Appl Genet 2003; 106(4) 594–598.

[67] James AC, Peraza-Echeverria S, Herrera-Valencia V, Martinez O. Application of the amplified length polymorphism (AFLP) and the methylation-sensitive amplification polymorphism (MSAP) techniques for the detection of DNA polymorphism and changes in DNA methylation in micropropagated bananas. In: Banana improvement: Cellular, Molecular Biology, and Induced Mutations, ed. Jain SM, Swennen R. Science Publishers, Enfield (NH) USA & Plymouth, UK, 2004, p 97-110.

[68] Vroh-Bi I, Anagbogu C, Nnadi S, Tenkouano A. Genomic characterization of natural and somaclonal variations in bananas (*Musa* spp.). Plant Mol Biol Rep 2011; 29(2) 440–448.

[69] Youssef M, Ku-Cauich R, James A, Escobedo-Gracia Medrano RM. Genetic analysis of somatic embryogenesis derived plants in banana. Assiut Journal of Agricultural Sciences   2011;   42:   287-300.http://www.researchgate.net/publication/ 237101174_Genetic_Analysis_of_Somatic_Embryogenesis_Derived_Plants_in_Banana/file/72e7e51b893e19bb12.pdf (accessed 17 November 2014).

[70] Buetow KH, Edmonson MN, Cassidy AB. 1999. Reliable identification of large numbers of candidate SNPs from public EST data. Nat Genet 21: 323–325. http:// www.aun.edu.eg/faculty_agriculture/arabic/journal/english/w_elkhorashy.pdf (accesed 20 October 2014).

[71] Soleimani VD, Baum BR, Johnson DA. Efficient validation of single nucleotide polymorphisms in plants by allele-specific PCR, with an example from barley. Plant Mol Biol Rep 2003; 21(3) 281–288.

[72] Mammadov J, Aggarwal V, Buyyarapu R, Kumpatl S. SNP Markers and their impact on plant breeding. International Journal of Plant Genomics 2012: 728398. http:// dx.doi.org/10.1155/2012/728398 (accesed 20 October 2014).

[73] Wenzl P, Carling J, Kudrna D, Jaccoud D, Huttner E, Kleinhofs A, Kilian A. Diversity arrays technology (DArT) for whole-genome profiling of barley. Proc Natl Acad Sci USA 2004; 101(26) 9915-9920.

[74] Kilian A. Towards effective deployment of diversity arrays technology (DArT) in banana genomics and sequencing. Conference Proceedings January 13–17, 2007, Plant and Animal Genomes XV Conference. San Diego, USA. 2007

[75] Amorim EP, Vilarinhos AD, Cohen KO, Amorim VBO, Dos Santos-Serejo JA, Silva SO, Pestana KN, Dos Santos VJ, Paes NS, Monte DC, Dos Reis RV. Genetic diversity of carotenoid-rich bananas evaluated by Diversity Arrays Technology (DArT). Genetics and Molecular Biology 2009; 32(1) 96-103.

[76] Risterucci AM, Hippolyte I, Perrier X, Xia L, Caig V, Evers M, Huttner E, Kilian A, Glaszmann JC. Development and assessment of diversity arrays technology for high-throughput DNA analyses in *Musa*. Theor Appl Genet 2009; 119(6) 1093-1103.

[77] Sales EK, Butardo NG, Paniagua HG, Jansen H, Dolezel J. 2001. Assessment of ploidy and genome constitution of some *Musa balbisiana* cultivars using DArT Markers. Philippine Journal of Crop Science 2011; 36(1) 11-18.

[78] Mbanjo EGN, Tchoumbougnang F, Mouelle AS, Oben JE, Nyine M, Dochez C, Ferguson ME, Lorenzen J. Molecular marker-based genetic linkage map of a diploid banana population (*Musa acuminata* Colla). Euphytica 2012; 188:(3) 369-386.

[79] Colbert T, Till BJ, Tompa R, Reynolds S, Steine MN, Yeung AT, McCallum CM, Comai L, Henikoff S. High-throughput screening for induced point mutations. Plant Physiol 2001; 126(2) 480–484.

[80] Till BJ, Zerr T, Bowers E, Greene EA, Comai L, Henikoff S. High-throughput discovery of rare human nucleotide polymorphisms by Ecotilling. Nucleic Acids Res 2006; 34(13) e99.

[81] Comai L, Young K, Till BJ, Reynolds SH, Greene EA, Codomo CA, Enns LC, Johnson JE, Burtner C, Odden AR, Henikoff S. Efficient discovery of DNA polymorphisms in natural populations by Ecotilling. Plant J 2004; 37 (5) 778–786.

[82] Mmeka EC, Adesoye AI, Vroh BI, Ubaoji KI. 2013. Single nucleotide polymorphism (SNP) markers discovery within *Musa* spp (plantain landraces, AAB genome) for use in beta carotene (Provitamin A) trait mapping. American Journal of Biology and Life Science 2013; 1 (1) 11-19.

[83] Mahendhirana M, Ramirez-Prado JH, Escobedo-Gracia Medrano RM, Canto-Canché B, Tzec-Simá M, Grijalva-Arango R, James-Kay A. Single nucleotide polymorphisms in partial sequences of the gene encoding the large sub-units of ADP-glucose pyrophosphorylase within a representative collection of 10 *Musa* genotypes. Electronic Journal of Biotechnology 2014; 17 (3) 137–147.

[84] Yunis JJ. 1976. High resolution of human chromosomes. Science 191(4233) 1268-1270.

[85] Doležalová M, Valárik M, Swennen R, Horry JP, Doležel J. Physical mapping of the 18S-25S and 5S ribosomal RNA genes in diploid bananas. Biol Plantarum 1998; 41(4) 497-505.

[86] Osuji JO, Crouch J, Harrison G, Heslop-Harrison JS. Molecular cytogenetics of *Musa* species, cultivars and hybrids: Location of 18S-5.8S-25S and 5S rDNA and telomere-like sequences. Annals of Botany 1998; 82(2) 243-248.

[87] Harper G, Osuji JO, Heslop-Harrison JS, Hull R. Integration of banana streak badna-virus into the *Musa* genome: Molecular and cytogenetic evidence. Virology 1999; 255(2) 207-213.

[88] Balint-Kurti PJ, Clendennen SK, Dolezelova M, Valarik M, Dolezel J, Beetham PR. Identification and chromosomal localization of the monkey retrotransposon in *Musa* sp. Mol Gen Genet 2000; 263(6) 908-915.

[89] Hřibová E, Doležalová M, Doležel J. Localization of BAC clones on mitotic chromo-somes of *Musa acuminate* using fluorescence *in situ* hybridization. Biologia Plantarum 2008; 52 (3) 445-452.

[90] De Capdeville G, Souza Junior MT, Szinay D, Diniz L, Wijnker E, de Jong H, Swen-nen R, Kema GHJ, De Jong H. The potential of high-resolution BAC-FISH in banana breeding. Euphytica 2009; 166 (3) 431-443.

[91] Neumann P, Navrátilová A, Koblížková A, Kejnovský E, Hřibová E, Hobza R, Widmer A, Doležel J, Macas J. Plant centromeric retrotransposons: A structural and cytogenetic perspective. Mobile DNA 2011; 2: 4. http://www.mobilednajournal.com/content/2/1/4 (accesed 20 October 2014).

[92] Cızkova J, Hribova E, Humplıkova L, Christelova P, Suchankova P, Dolezel J. Molec-ular analysis and genomic organization of major DNA satellites in banana (*Musa* spp.). PLoS ONE 2013; 8(1) e54808. http://www.plosone.org/article/info%3Adoi%2F10.1371%2Fjournal.pone.0054808 (accesed 20 October 2014).

[93] Schwarzacher T, Leitch AR, Bennett MD, Heslop-Harrison JS. *In situ* localization of parental genomes in a wide hybrid. Annals of Botany 1989; 64(3) 315–324.

[94] Osuji JO, Crouch J, Harrison G, Heslop-Harrison JS. Identification of the genomic constitution of *Musa* L. lines (bananas, plantains and hybrids) using molecular cyto-genetics. Annals of Botany 1997; 80(6) 787–793.

[95] D'Hont A, Paget-Goy A, Escoute J, Carreel F. The interspecific genome structure of cultivated banana, *Musa* spp. revealed by genomic DNA *in situ* hybridization. Theo-retical and Applied Genetics 2000; 100(2) 177-183.

[96] D'Hont A. Unraveling the genome structure of polyploids using FISH and GISH; ex-amples of sugarcane and banana. Cytogenet Genome Res 2005; 109 (1-3) 27-33.

[97] Jeridi M, Bakry F, Escoute J, Fondi E, Carreel F, Ferchichi A, D'Hont A, Rodier-Goud
      M. Homoeologous chromosome pairing between the A and B genomes of *Musa* spp.
      revealed by genomic in situ hybridization. Annals of Botany 2011; 108(5) 975–981.

[98] Doležel J, Doleželová M, Novák F. Flow cytometric estimation of nuclear DNA
      amount in diploid bananas (*Musa acuminata* and *M. balbisiana*). Biologia Plantarum
      1994; 36(3) 351–357.

[99] Kamaté K, Brown S, Durand P, Bureau JM, De Nay D, Trinh TH. Nuclear DNA con-
      tent and base composition in 28 taxa of *Musa*. Genome 2001; 44(4) 622–627.

[100] Asif MJ, Mak C, Othman RY. Characterization of indigenous *Musa* species based on
      flow cytometric analysis of ploidy and nuclear DNA content. Caryologia 2001; 54(2)
      161–164.

[101] Bartos J, Alkhimova O, Doleželová M, De Langhe E, Doležel J. Nuclear genome size
      and genomic distribution of ribosomal DNA in *Musa* and *Ensete* (Musaceae): Taxo-
      nomic implications. Cytogenet Genome Res 2005; 109(1-3) 50-57.

[102] Lysak MA, Doleželová M, Horry JP, Sweenen R, Doležel. Flow cytometry analysis of
      nuclear DNA content in *Musa*. Theor Appl Genet 1999; 98(8) 1344–1350.

[103] De Langhe E, Hřibová E, Carpentier S, Doležel J, Swennen R. Did backcrossing con-
      tribute to the origin of hybrid edible bananas? Annals of Botany 2010; 106(6) 849–857.

[104] de Jesus ON, Silva SdO, Amorim EP, Ferreira CF, de Campos JMS, de Gaspari GS,
      Figueira A. Genetic diversity and population structure of *Musa* accessions in ex situ
      conservation. BMC Plant Biology 2013; 13: 41. http://www.biomedcentral.com/
      1471-2229/13/41 (accesed 20 October 2014).

[105] Roux N, Strosse H, Toloza A, Panis B, Dolezel J. Detecting ploidy level instability of
      banana embryogenic suspension cultures by flow cytometry. In: Jain MS, Swennen R.
      (eds.) Banana Improvement: Cellular, Molecular, Biology, and Induced Mutations In
      Sci Publishers Inc., 2004. p251–261.

[106] Roux N, Toloza A, Radeck Z, Zapata-Arias FJ, Doležel J. Rapid detection of aneuploi-
      dy in *Musa* using flow cytometry. Plant Cell Reports 2003; 21(5) 483–490.

[107] Escobedo-Gracia Medrano RM, Maldonado-Borges J, Burgos-Tan M, Valadez-Gonzá-
      lez N, Ku-Cauich JR. Using flow cytometry and cytological analyses to assess the ge-
      netic stability of somatic embryo-derived plantlets from embryogenic *Musa acuminata*
      Colla (AA) ssp. *malaccensis* cell suspension cultures. Plant Cell, Tissue and Organ
      Culture (PCTOC) 2014; 116(2) 175-185.

[108] Mueller UG, Wolfenbarger LL. AFLP genotyping and fingerprinting. TREE 1999;
      14(10) 389-394.

[109] Carpentier SC, Witters E, Laukens K, Van Onckelen H, Swennen R, Panis B. Banana (*Musa* spp.) as a model to study the meristem proteome: Acclimation to osmotic stress. Proteomics 2007; 7(1) 92–105.

[110] Yang QS, Wu JH, Li CY, Wei YR, Sheng O, Hu CH, Kuang RB, Huang YH, Peng XX, McCardle JA, Chen W, Yang Y, Rose JK, Zhang S, Yi GJ. Quantitative proteomic analysis reveals that antioxidation mechanisms contribute to cold tolerance in plantain (*Musa paradisiaca* L.; ABB Group) seedlings. Mol Cell Proteomics 2012; 11(12) 853–869.

[111] Carpentier SC, Panis B, Renaut J, Samyn B, Vertommen A, Vanhove AC, Swennen R, Sergeant K. The use of 2D-electrophoresis and de novo sequencing to characterize inter-and intra-cultivar protein polymorphisms in an allopolyploid crop. Phytochemestry 2011; 72(10)1243-1250.

[112] Toledo TT, Nogueira SB, Cordenunsi BR, Gozzo FC, Pilau EJ, Lajolo FM, Oliveira do Nascimento JR. Proteomic analysis of banana fruit reveals proteins that are differentially accumulated during ripening. Postharvest Biology and Technology 2012; 70: 51-58. http://www.sciencedirect.com/science/article/pii/S0925521412000816 (accesed 20 October 2014).

[113] Esteve C, D'Amato A, Marina ML, García MC, Righetti PG. In-depth proteomic analysis of banana (*Musa* spp.) fruit with combinatorial peptide ligand libraries. Electrophoresis 2013; 34(2) 207-214.

[114] Li X, Bai T, Li Y, Ruan X, Li H. 2013. Proteomic analysis of *Fusarium oxysporum* f. sp. *cubense* tropical race 4-inoculated response to Fusarium wilts in the banana root cells. Proteome Science 2013; 11: 41 http://www.proteomesci.com/content/11/1/41 (accesed 20 October 2014).

7

Biodiversity and Plant Breeding as Tools for Harmony Between Modern Agriculture Production and the Environment

João Carlos da Silva Dias

1. Introduction

There are now almost 7.25 billion human beings inhabiting this planet, and it has been projected that world population growth may exceed 70 million annually over the next 40 years. The world population will be approximately 9.2 billion in 2050, when the concentration of carbon dioxide and ozone will be 550 ppm and 60 ppm, respectively and the climate will be warmer by 2ºC [1]. At that time it is expected that approximately 90% of this global population will reside in Asia, Africa, and Latin American countries [2,3]. Currently, about 1 billion human beings suffer from hunger; 3 billion malnourished people suffer one or more micronutrient deficiencies (especially vitamin A, iodine and iron) and live with less than 2 US dollars per day; and anthropogenic climate change continues to affect food output and quality [4,5]. By 2050, to sufficiently feed all these people, the total food production will have to increase 60 to 70% to meet a net demand of 1 billion tonnes of cereal for food and to feed, and 200 million tonnes of meat [6-8], depending on assumptions of population growth, income growth and dietary changes. This projected increase of global crop demand is partly due to a growing global population, but a larger driver is increasing global affluence and associated changes in diet due to higher incomes [4,8]. As global incomes increase, diets typically shift from those comprised of mostly grains, to diets that contain greater proportion of meat, dairy products, and eggs and more vegetables and fruits [4,8-10].

In order to meet these demands, global livestock production systems are shifting from using mostly marginal lands and crop residues to more industrial systems which require less land and use of higher value feed crops [11,12]. Increasing demand for meat and dairy products is also of importance to the global environment because their production requires more land, water and other resources [13-15]. Livestock production is also responsible for other environmental impacts. Besides livestock production is estimated to be responsible for 18% of total greenhouse gas emissions [16], and animal products generally have a much higher water footprint than vegetal products [17].

In 2008, the world's arable land amounted to 1,386 M ha, out of a total 4,883 M ha land used for agriculture [18]. Each year, arable and agricultural land is lost due to deforestation, overgrazing, agricultural activities, gathering and overexploitation for fuel-wood, urbanization and industrialization. The most direct negative impact of agriculture on biodiversity is due to the considerable loss of natural habitats, which is caused by the conversion of natural ecosystems into agricultural land. The arable land is limited. Increases in arable land can only be done by deforestation. Agricultural production should be increased without further deforestation. This requires innovation and better technologies, as well as substantial investment, to increase yields on existing agricultural land.

Climate models predict that warmer temperatures and increases in the frequency and duration of drought during the twenty-first century will have negative impact on agricultural productivity [19-24]. For example, maize production in Africa could be at risk of significant yield losses as researchers predict that each degree-day that the crop spends above 30°C reduces yields by 1% if the plants receive sufficient water [23]. These predictions are similar to those reported for maize yield in the United States [25]. Lobell et al. [23] further showed that maize yields in Africa decreased by 1.7% for each degree-day the crop spent at temperatures of over 30°C under drought. Wheat production in Russia decreased by almost one-third in 2010, largely due to the summer heat wave. Similarly, wheat production declined significantly in China and India in 2010, largely due to drought and sudden rise in temperature respectively, thereby causing forced maturity [26]. Warming at +2°C is predicted to reduce yield losses by 50% in Australia and India [27,28]. Likewise, the global maize and wheat production, as a result of warming temperatures during the period of 1980 to 2008, declined by 3.8% and 5.5%, respectively [24]. So climate change poses a serious threat to species fitness [29,30], and to agro-ecosystems essential to food production [31].

Climatic variation and change are already influencing the distribution and virulence of crop pest and diseases, but the interactions between the crops, pests and pathogens are complex and poorly understood in the context of climate change [32]. We will need to integrate plant biology into the current paradigm with respect to climate change to succeed in defeating emerging pests and pathogens posing a new threat to agriculture due to climate change [33-35].

In this context we can ask: can we feed and clothe the growing world population while simultaneously preserving or improving ecosystems and the natural environment?

History shows that modern agriculture has the potential to feed the world population but also to be worst and even catastrophically with the natural environment. Some examples are deforestation, overgrazing and erosion, in many parts of the world, which contributed to the outright collapse of ecosystems. One classical example is Madagascar's central highland plateau that has become virtually totally barren (about ten percent of the country), as a result of slash-and-burn deforestation, an element of shifting cultivation practiced by many natives.

Intensification of production systems have also led to reduction in crop and livestock biodiversity, and increased genetic vulnerability and erosion. In contrast, the "Green Revolution", which began providing high-yielding crop cultivars and high-input management techniques to developing countries in the 1960s, has prevented mass starvation and improved living standards throughout the world [36]. Dwarfing, photoperiod insensitive genes and host plant resistance genes to pathogens and pests were bred for various crops during the "Green Revolution" [37]. Crop yields were increased in many nations of Asia and Latin America by innovations of the "Green Revolution". Calorie consumption would have dropped by about 5% and the number of malnourished children would have increasing by at least 2%; i.e., the "Green Revolution" helped to improve the health status of 32 to 42 million pre-school children. Since the beginning of the "Green Revolution" in 1960, land devoted to crops increased some 10%, land under irrigation has doubled, pesticide use by agriculture has tripled, fertilizer use is up 23-fold, pesticide use is up by a factor of 53. Nowadays, forty per cent of crop production comes from the 16% of agricultural land that is irrigated. Irrigated lands account for a substantial portion of increased yields obtained during the "Green Revolution". The enhancement of yield achieved in the "Green Revolution" (29% in food supplies per capita since 1960) may have been associated with an increased level of greenhouse gas emissions associated with higher fertilizer production and application, but, overall, its net effect has been calculated to have reduced CO_2 emission by some 161 gigatons of carbon (GtC) over the period 1961-2005 [38], implying that gains in crop productivity can make a positive contribution to reducing greenhouse gas emissions.

Developing sustainable agriculture in environmentally sensitive systems is the great challenge of the coming decades. More food, animal feed, fiber, fuel, and forest products must be produced with less available land, water, and nutrients, to meet basic human needs and improve the sustainability of production [39]. In addition, pressure from an increasing global human population will necessitate more efficient and diversified land use.

Identifying the most appropriate technologies and practices to achieve these objectives are critical. This requires the building of a knowledge base to support such tasks. Agro-ecological approaches are known to increase farming system productivity, to reduce pollution, and to maintain biodiversity through careful management of soil, water, and natural vegetation. The agenda for a new "Green Revolution" needs to consider new approaches to promote innovations in plant science, agricultural and management practices and benefits to farmers and consumers.

Modern production agriculture in the developed world is highly industrialized. There is considerable discussion about the inadequacy of the dominant model of agricultural intensification and growth, which relies on increased use of capital inputs, such as fertilizer and pesticides [40]. Technology and purchased inputs, e.g. fertilizer, pesticides and water are required to maintain high levels of production, and use of these inputs continues to increase in the developing world. Despite the critical need for agricultural production and continued improvements in management practices, current systems are still not in "harmony" with the environment because they can create many problems for ecosystems and human communities. The generation of unacceptable levels of environmental damage and problems of economic feasibility are cited as key problems with this model of industrial agriculture [39,41]. Specific external costs of industrial agriculture which should be improved include soil deterioration, erosion, declining surface water and groundwater quality, limited recycling of nutrients, excessive use of off-farm fertilizers and pesticides, diminished biodiversity within the agricultural system (both in terms of the variety of crops sown and coexisting species), lapses in food safety, and the loss of rural employment. By developing new field crops, and trees that meet societal needs, plant breeding plays a distinctive and crucial role in addressing these challenges, which must be dealt with immediately to develop sustainable agronomic systems for the future.

In this article two general ways are described in which plant breeders can engage in environmental issues: i) by breeding plants that are better adapted to environment and environmental stresses, producing more with less and where productivity can be maintained in the face of increasingly variable weather patterns and sub-optimal conditions, as well as pest and disease pressures; and ii) by breeding plants that can alter and "improve" environments, as breeding alternative crops and crops for new uses or breeding for local adaptation and sustainable solutions. Previously, the concepts of crop biodiversity, soil biodiversity and agro-biodiversity were briefly presented.

2. Crop biodiversity, soil biodiversity and agro-biodiversity

2.1. Crop biodiversity

Today, 150 plant species (out of 250,000 known plant species) dominate the world's agricultural landscapes, but only 12 crop species provide 80% of the world's food chain [42]. Three main cereals: wheat, rice and maize, provide about 50% of the energy we obtain from plants.

The wise use of crop genetic diversity in plant breeding can contribute significantly to protect the environment. A major role of genetic resources will be to provide germplasm resistant to pests and diseases, more efficient in their use of water and nutrients and less dependent on external inputs to maintain current levels of productivity. Natural genetic diversity is becoming increasingly important to understanding the ways in which we can improve plant breeding. There is a continuing need to assemble and screen germplasm strategically and discover new sources of variation that will enable developing new crop cultivars. Complex traits can be improved dramatically by bringing novel alleles from diverse ecotypes into breeding material.

Crop genetic biodiversity is considered a source of continuing advances in yield, disease and pest resistance, and quality improvement. It is widely accepted that greater varietal and species diversity would enable agricultural systems to maintain productivity over a wide range of conditions. The loss of biodiversity is considered one of today's most serious environmental concerns. In the last 50 years vegetable genetic resources have been lost, on a global scale at the rate of 1-2% per year [43] and it has been estimated by FAO that 6% of wild relatives of cereal crops (wheat, maize, rice, etc.) are under threat as well as 18% of legume species, and 13% of solanaceous [44].

There is a growing world-wide awareness about the need to conserve plant germplasm for the use of future generations. Consequently, considerable media attention has been given to the creation of the Svalbard Global Seed Vault (see http://www.croptrust.org/main) and relates to storage of seeds of many economically important crops [3,45]. Gene banks are crop genetic diversity reservoirs and sources of alleles for sustainable genetic enhancement of crops [46]. Indeed breeding gains depend on capitalizing on the useful genetic variation present in the crop gene pools, which for many crops is being conserved in gene banks. There are about 1,700 gene banks and germplasm collections around the world (the number in FAO's database). They maintain about 7.4 million accessions of plant genetic resources, with cereals and legumes constituting 52% of the accessions [47]. The CGIAR consortium holds about 0.7 million accessions of 3,446 species from 612 genera. The International Crops Research Institute for the Semi-arid Tropics (ICRISAT) possesses one of the largest gene banks in the world with approximately 115,000 accessions of cereals (sorghum, millets) and legumes (chick-pea, groundnut, pigeon-pea) [48]. In spite of these large collections maintained *ex situ*, there are still important collection gaps that must be addressed before these priceless genetic resources are lost as a result of climate change or other driving forces leading to the genetic erosion and loss of biodiversity [47,49]. These *ex situ* collections are to a large extent safe from the adverse impact of climate change.

Ex situ collections should be subjected to phenotypic, disease resistance and molecular characterization to facilitate the potential use of this genetic endowment for the amelioration of crops. Plant breeders seldom access accessions from some gene banks with large collections. A systematic assessment of the genetic diversity in such collections has helped to establish core collections, which should be subsets of large collections [50-52], containing chosen accessions that capture most of the genetic variability in the entire collection. A core collection therefore improves the management and utilization of a germplasm collection. Genetic studies in selected crops have shown that both common widespread and localized alleles occurring in the entire collection are contained in the core collection subset. Only rare localized alleles may be excluded during the aforementioned sampling process. The core collection subset often provides an entry point to the entire collection for further investigation of the genetic diversity or for the utilization of these resources. Core collections which are *a priori* selected by the curator are often of limited use to those users of the gene bank germplasm collection who are interested in specific trait or domain. The current revolution in information technology makes it possible for users to make such selections themselves directly on the Worldwide Web using a stratified sampling in the domain(s) of interest. This approach allows a more focused selection of the germplasm accessions which shows variation for the trait of interest to the user compared with the use of core collections. These smaller core collections are sometimes enough to capture most of the useful variations.

Research undertaken on the large global collection of sorghum landraces and genetic stocks held at ICRISAT (in excess of 35,000) demonstrates how the challenge of maintaining a large number of accessions and the related information documented for this collection can be addressed by gene bank curators. Different sampling strategies were proposed to obtain core collection subsets of reduced size [53]. Three core collections subsets were established following: i) a random sampling within a stratified collection (logarithmic strategy); ii) non-random sampling based upon morpho-agronomic diversity (principal component score strategy); and non-random sampling based upon an empirical knowledge of sorghum (taxonomic strategy). These core collections subsets did not differ significantly in their overall phenotypic diversity according to principal component representation of the morpho-agronomic diversity using the Shannon-Weaver diversity indice. But when comparisons for morpho-agronomic diversity and passport data were considered, the principal component strategy subset looked similar to the entire landrace collection. The logarithmic strategy subset showed differences for characters associated with the photoperiod reaction that was considered in the random sampling stratification of the collection. The taxonomic strategy subset was the most distinct subset from the entire landrace collection. It represented the landraces selected by farmers for specific uses and covered the widest range of geographical adaptation and morpho-agronomic traits.

In the same sorghum landraces collection of ICRISAT, partial assessment of host response to five sorghum diseases provided another means to quantify the importance of agro-biodiversity in resistance [54]. Frequency distributions of host response to major sorghum pathogens were the same between the entire collection and core collection subsets for all diseases, except between the entire collection and the logarithmic core subset for grain mold. This was not surprising because the sampling strategy for this core subset and the material included in the screening for this disease did not match. The logarithmic core subset had the widest range of adaptation to photoperiod whereas only photoperiod insensitive germplasm had been screened for grain mold. The lack of accessions that fall in the highest resistance class for some diseases in the core subsets is the result of sampling statistics, but the χ^2 tests for homogeneity clearly confirmed that the entire collection and the core subsets included the same distribution of variation with only the above stated exception for grain mold in the logarithmic strategy core subset. New accessions with high resistance to specific diseases are likely to be identified by completing the screening of the core subsets. This rational, targeted approach may also be cost-effective and more precise than long term screening of the entire collection. Furthermore this analysis also shows that large sample sizes do not appear to always be associated with capturing useful variation for disease resistance (i.e., entire *vs.* core collections), neither when the sampling was defined by breeding objectives (like the logarithmic strategy subset), a mirror of the entire collection (principal component score strategy subset) or by maximizing farmer's landraces (taxonomic strategy subset).

The latest database on world plant genetic resources highlighted that there are still large gaps, more specifically in crop wild relatives and landraces, in *ex situ* gene bank collections preserved across the globe [55]. Unlike cultivated germplasm, there are difficulties associated with *ex situ* conservation of crop wild relatives due to their specific crop husbandry, tendency for natural pod dehiscence, seed shattering and seed dormancy, high variability in flowering and seed production, and rhizomatous nature of some of the species. Crop wild relatives have contributed many agronomically beneficial traits in shaping the modern cultivars [56], and they will continue to provide useful genetic variations for climate-change adaptation, and also enable crop genetic enhancers to select plants which will be well-suited for the future's environmental conditions [57]. There is a growing interest that crop wild relatives should be preserved *in situ* in protected areas to ensure the evolutionary process of wild species contributing new variants, which as and when captured by plant explorers, should be able to contribute to addressing new challenges to agricultural production [58]. Worldwide, there are 76,000 protected areas, spread in ~17 million km^2, and several countries have taken initiatives to establish crop wild relative's *in situ* conservation [59-61]. Promoting *in situ* conservation may allow genes to evolve and respond to new environments that would be of great help to capture new genetic variants helping to mitigate climate-change impacts [62].

2.2. Soil biodiversity

Biodiversity and soil are strongly linked, because soil is the medium for a large variety of organisms, and interacts closely with the wider biosphere. Soil biodiversity exceeds the aboveground systems biodiversity, and is crucial for the sustainability of agro-ecosystems [63]. It consists of macrofauna or soil engineers (earthworms and termites), mesofauna (microarthropods such as mites and springtails), microfauna (nematodes and protozoans), and microflora (bacteria and fungi). The soil organisms perform a number of vital functions such as: i) decomposition and degradation of plant litter and cycling of nutrients; ii) converting atmospheric nitrogen into organic forms (immobilization) and remineralization of mineral nitrogen, leading to the formation of gaseous nitrogen; iii) suppression of soil pathogens through antagonism; iv) regulating microclimate and local hydrological processes; v) synthesizing enzymes, vitamins, hormones, vital chelators and allelochemicals that regulate population and processes; vi) altering soil structure and other soil physical, chemical and biological characteristics; and vii) microbial exudates have a dominant role in the aggregation of soil particles and the protection of carbon from further degradation [64,65]. Biological activity helps in the maintenance of relatively open soil structure; it facilitates decomposition and its transportation as well as transformation of soil nutrients. It is not surprising that soil management has a direct impact on biodiversity. This includes practices that influence global changes, soil structure, biological and chemical characteristics, and whether soil exhibits adverse effects such as soil acidification.

Soil acidification has an impact on soil biodiversity. Roem and Berendse [66] in the Netherlands, examined the correlation between soil pH and soil biodiversity in soils with pH below 5 in grassland and heath land communities. A strong correlation was discovered, wherein the lower the pH the lower the biodiversity. Soil acidification reduced the numbers of most macrofauna and affected rhizobium survival and persistence. So extremely low pH soils may suffer from structural decline as a result of reduced microorganisms. This brings a susceptibility to erosion under high rainfall events, drought, and agricultural disturbance.

Land use pattern, plant diversity, soil desertification and pollution, including those resulting from N enrichment, alter soil biodiversity [67-69]. The changes in soil biodiversity are also observed through effects on soil organisms as a result of the changes in temperature and precipitation and through climate-driven changes [like rising atmospheric/ambient CO_2 (hereafter aCO_2) and warming] in plant productivity and species composition.

Accumulated evidence so far reveals that soil biota is vulnerable to global changes and soil disturbance. Castro et al. [70], in a multifactor climate change experiment, reported increased fungal abundance in warmed treatments, increased bacterial abundance in warmed plots with elevated atmospheric CO2 (hereafter eCO_2) but decreased in warmed plots under aCO_2, changes in precipitation altered the relative abundance of proteobacteria and acidobacteria where acidobacteria decreased with a concomitant increase in the proteobacteria in wet relative to dry treatments, altered fungal community composition due to the changes in precipitation, and differences in relative abundance of bacterial and fungal clones varied among treatments. All these observations led the researchers to conclude that climate change drivers and their interactions among them may cause changes in the bacterial and fungal abundance, with precipitation having greater effect on the community composition.

Dominique et al. [71] in their research, where the influence of plant diversity and eCO_2 levels on belowground bacterial diversity were analyzed observed that the variability in plant diversity level had significant effects on bacterial composition but no influence on bacterial richness. This research therefore suggests that the soil microbial composition is mainly related to plant diversity, assuming that different plant species might harbor specific rhizospheric microbial populations, rather than altered soil carbon fluxes induced by eCO_2 which can lead to increased photosynthesis. Bardgett [72] points out there is sufficient evidence to show that the transfer of carbon through plant roots to the soil plays a primary role in regulating ecosystem responses to climate change and its mitigation.

Very little is known about the influence of eCO_2 on the structure and functioning of below ground microbial community. In a 10-year field exposure of a grassland ecosystem to eCO_2, Zhili et al. [73] detected a dramatic alteration in the structure and functional properties of soil microbial communities. They found the total microbial and bacterial biomass significantly increased under eCO_2, while the fungal biomass remained unaffected. Furthermore, the structure of microbial communities was markedly different between aCO_2 and eCO_2. More recently, using tag-encoded pyrosequencing of 16S rRNA genes, Deng et al. [74] also found that the soil microbial community composition and its structure were significantly altered under eCO_2. In both studies, the changes in microbial structure were significantly correlated to soil moisture, soil status relative to C and N contents, and plant productivity.

2.3. Agro-biodiversity

Agro-biodiversity is the result of the interaction between the environment, the variety and variability of animals, plants and microorganisms that are necessary for sustaining key functions of the agro-ecosystem, and the management systems and practices. It is the human activity of agriculture which shapes and conserves this biodiversity.

Agro-biodiversity consists of the genetic diversity within the species, the species diversity, and the ecosystem diversity, which comprises the variation between agro-ecosystems within a region.

There are several distinctive features of agro-biodiversity, compared to other components of biodiversity: i) agro-biodiversity is actively managed by farmers and would not survive without this human interference; ii) due to the degree of the human management and inter-ference, conservation of agro-biodiversity in production systems is inherently linked to sustainable use; iii) many economically important agricultural systems are based on 'alien' livestock and crop species introduced from elsewhere; iv) in regards to crop diversity, diversity within species is at least as important as diversity between species; and v) as stated before in industrial-type agricultural systems, much crop diversity is now held *ex situ* in gene banks or breeders' materials rather than on-farm.

Agro-biodiversity provides the main raw material for intensifying sustainable crop yields and for adapting crops to climate change, because it can provide traits for plant breeders and farmers to select input-efficient, resilient, climate-ready crop germplasm and further release of new cultivars. Agro-biodiversity is crucial to cope with climate changes as the entire diversity of genes, species and ecosystems in agriculture represents the resource base for food [58]. Many farmers, especially those in environments where high-yield crop cultivars and livestock races do not prosper, rely on a wide range of crop and livestock types. This is the best method for increasing the reliability of food production in the face of seasonal variation. Diversified agricultural systems not only render smallholder farming more sustainable, but also reduce the vulnerability of poor farmers since they can minimize the risk of harvest failures caused by the outbreak of diseases and pests, by droughts or floods, or by extremely high temperatures, all of which will be exacerbated by climate change [75].

Monoculture means growing a single plant species in one area. Monoculture however should not be regarded as synonymous to a single crop cultivar in a farmers' field since monoculture can present intra-specific genetic diversity. For instance, a crop under monoculture can be a mixture of distinct cultivars or landraces having genetic variation within each population. Intra-specific crop diversification can provide a means of effectively controlling diseases and pests over large areas and therefore contribute to sustainable intensification of crop production. Nonetheless, an agro-ecosystem with many species of different taxa will be richer in species diversity than another agro-ecosystem where many species of the same taxon occur. Genetically diverse populations and species-rich agro-ecosystems may show greater buffer potential to adapt to climate change. Agro-biodiversity at the gene, species and agro-ecosystem levels increase resilience to the changing climate. Promoting agro-biodiversity remains therefore crucial for resilience of agro-ecosystems.

There is much evidence that global agriculture would benefit from an intensified utilization of existing biodiversity. We need to shift the focus of agricultural research from genes alone to management and their interactions. There is much to be gained with mixed cropping, as shown in a study performed by Tilman et al. [76], where plots with 16 species produced 2.7 times more biomass than monocultures. Bullock et al. [77] in comparing meadows with different number of species, found, after 8 years experiment, that the richer meadows yielded 43 % more hay than species-poor fields. Increased grassland diversity promotes temporal stability at many levels of ecosystem organization [78]. Mixtures of barley cultivars in Poland generally out-yielded the means of cultivars as pure stands [79]. The highly intensive agricultural system of home gardens are some of the most diverse production systems in the world and also some of the most productive [80,81]. Agro-biodiversity in home gardens reduces year-to-year variation, thus contributing to stability in yield. Although they are usually highly labor intensive and small, they provide income and nutrition for millions of small farmers throughout the world.

3. Plant breeding, agriculture and environment

3.1. Introduction

Farming and plant breeding have been closely associated since the early days when crops were first domesticated. The domestication of staple crops, for example, rice and soybean in eastern Asia; wheat in the Middle East; sorghum in Africa; and maize, beans, and potatoes in the Americas [82], began independently, in multiple locales, 5000-12 000 years ago [82]. For thousands of years, these crops were grown and morphologically altered by farmers, who selected the most desirable and adaptable cultivars to plant in the next growing season. Without understanding the science behind it, early farmers saved the seed from the best portion of their crop each season. Over the years, they selected the traits which they liked best, transforming and domesticating the crops they grew.

After the discoveries of Darwin and Mendel, scientific knowledge was applied to plant breeding in the late 1800s [36]. Commercial hybridization of crop species began in the United States in the middle of the 1920s with sweet corn and followed by onions in the 1940s [4]. With the implementation of hybrid crop breeding, yield per unit land area rapidly increased in the United States [83] and since that time, public and private breeding companies have been placing more and more emphasis on the development of hybrids, and many species have been bred as hybrid cultivars for the marketplace. Besides heterosis, hybrids also allow breeders to combine the best traits and multiple disease and stress resistances. Furthermore, if the parents are homozygous, the hybrids will be uniform, an increasingly important trait in commercial market production. The creation of hybrid cultivars requires homozygous inbred parental lines, which provide a natural protection of plant breeders' rights without legal recourse and ensure a market for seed companies.

In the 1970's breeders' rights protection has been provided through International Union for the Protection of New Varieties of Plants (UPOV), which coordinates an international common legal regime for plant variety protection. Protection was granted for those who develop or discover cultivars that are new, distinct, uniform, and stable [84]. Cultivars may be either sexually or asexually propagated. Coverage for herbaceous species is 20 years. Protective ownership was extended by UPOV in 1991 to include essentially derived cultivars [84]. At the same time, the farmer's exemption (which permitted farmers to save seed for their own use) was restricted; giving member states the option to allow farmers to save seed. Additionally, in Europe after 1998 and the United States after 2001, plant breeding companies can take advantages of patent laws to protect not only the cultivar itself but all of the plant's parts (pollen, seeds), the progeny of the cultivar, the genes or genetic sequences involved, and the method by which the cultivar was developed [85]. The seed can only be used for research that does not include development of a commercial product i.e., another cultivar, unless licensed by the older patent. The patents are considered the ultimate protective device allowing neither a farmer's exemption nor a breeder's exemption (that permitted the protected cultivar to be used by others in further breeding to create new cultivars) [86]. The use of patents for transgenic crops introduces additional problems according to the IAASTD report [41] developed with the contribution from 400 scientists around the world, and adopted by 58 governments. In developing countries, especially instruments such as patents may boost up costs and restrict experimentation by individual farmers whereas potentially undermining local practices for securing food and economic sustainability. Thus, there is particular concern regarding present intellectual property rights instruments, which may inhibit seed-saving, exchange, sale, and access to proprietary materials of vital importance to the independent research community, specifically in view of the need for analyses and long term experimentation on climate change impacts [84,85].

Research and development (hereafter R&D) for improved seed development is expensive. Such product protection has presented a business incentive to corporations to invest in the seed industry, which supported an enormous increase in private R&D leading to strong competition in the marketplace between the major seed companies. The majority of current crop cultivars sold nowadays are proprietary products developed by private R&D. A significant consequence of this increase in R&D has been a reduction in public breeding programs. As a result, the cost for R&D to develop new crop cultivars is shifting from the publicly supported research programs to the customers of the major seed companies [4,87].

One of the main factors to determine success in plant breeding is crop biodiversity and genetic capacity. Access to genetic variation, biodiversity, is required to achieve crop cultivar improvement. No practical breeding program can succeed without large numbers of lines (genotypes) to evaluate, select, recombine and inbreed (fix genetically). This effort must be organized in order for valid conclusions to be reached and decisions to be made. Scientists, breeders, support people and facilities, budgets, and good management are requirements to assure success in the seed business. Science must be state-of-the-art to maximize success in a competitive business environment. The continued need for fundamental breeding research is critical to support development of new technology and expansion of the knowledge base which supports cultivar development, competition among proprietary cultivar results in owner-companies striving to do the best possible research to develop their own products and to compete on genetic and physiological quality of crop seed in the marketplace. Reasonable profit margins are essential to pay back the R&D costs to the owner and to fund future research on developing even better crop cultivars to stay competitive. There is considerable genetic variation within the numerous crop species, which can be exploited in the development of superior proprietary cultivars. The consequences of this dynamic situation will mean relatively short-lived cultivars replaced by either the owner of the cultivar or a competitor seed company. This intense competition means constantly improved and more sophisticated cultivars. Seed companies are in the business of manipulating genes to improve plant cultivar performance for a profit. The success of the research is judged by the success of the product in making a reasonable profit. The research must improve economic performance starting with the seed production costs and including the farmer-shipper/processor and the end user. If any link in this sequence of events is weak or broken, the new cultivar will likely fail [4,88].

Modern plant breeding is the science of improving plants to achieve farmer needs and better fit production environments, but it is a long-term proposition. Each released cultivar represents a culmination of a decade or more of work, from initial crosses through final testing. The rate of improvement is a function of the amount of heritable genetic variation present in a population, the time it takes to complete a breeding cycle (from seed production through selection to seed production again), which can range from multiple generations per year (e.g. maize on field sites in both hemispheres) to decades (some trees require 8 years of growth before flowering). In hybrid crops, several years (multiple breeding cycles) are necessary to develop inbred lines that must then be tested in hybrid combinations. Many years of testing under various environmental conditions must be conducted to ensure that the new cultivar (inbred, hybrid, or population) will perform well for the farmer, consumer, or end-user before any substantial additional investment is made to increase production and distribution of the cultivar.

Biotechnology is a new and potentially powerful tool that has been added by all the major seed corporations to their crop breeding research programs, and is part of an ongoing public research for developing genetic engineered crop projects. It can augment and/or accelerate conventional cultivar development programs through time saved, better products, and more genetic uniformity, or achieve results not possible by conventional breeding [89]. Genetic engineering provides innovative methods for modern plant breeding to adapt crops to agricultural systems facing new challenges brought by the changing climate. New breeding methods, relying on genetic engineering, can accelerate the pace to improve crops, or be more precise in transferring desired genes into plant germplasm. Some limited target traits already available in transgenic cultivars include those adapting agriculture to climate change and reducing their emissions of greenhouse gases.

Plant breeding may benefit from recent advances in genotyping and precise phenotyping, and by increasing the available agro-biodiversity through the use of genomics-led approaches. Today marker-assisted breeding is applied to a broad range of crops and could facilitate domesticating entirely new crops. Marker-assisted selection is particularly important for improving complex, quantitatively inherited traits that alter yield, and for speeding up the breeding process [90]. Crop genomics has also been improving in the last decade and today there are faster and cheaper systems being increasingly used in gene banks, genetic research and plant breeding, e.g. for studying interactions between loci and alleles such as heterosis, epistasis and pleiotropy, or analyzing genetic pathways. Advances in crop genomics are providing useful data and information for identifying DNA markers, which can be further used for both germplasm characterization and marker-assisted breeding. Genomics- assisted breeding approaches along with bioinformatics capacity and metabolomics resources are becoming essential components of crop improvement programs worldwide [84,91].

Progress in crop genome sequencing, high resolution genetic mapping and precise phenotyp-ing will accelerate the discovery of functional alleles and allelic variation associated with traits of interest for plant breeding. Genome sequencing and annotation include an increasing range of species such as wheat, rice, maize, sugarcane, potato, sorghum, soybean, banana, cassava, citrus, grape, among other species. Perhaps, one day further research on the genome of a plant species from a drought-prone environment may assist in breeding more hardy and water efficient related crops due to gene synteny.

Transgenic breeding involves the introduction of foreign DNA. While conventional plant breeding utilizing non-transgenic approaches will remain the backbone of crop improvement strategies, transgenic crop cultivars should not be excluded as products capable of contributing to development goals. Available commercial transgenic crops and products are at least as safe in terms of food safety as those ensuing from conventional plant breeding [89,92-94].

The use of transgenic crops remains controversial worldwide after almost two decades of introducing them into the agro-ecosystems. Using plant-derived genes to introduce useful traits and plant-derived promoters, may overcome some concerns about the development of genetically engineered crops. In this regard, cisgenesis addresses some negative views regarding the use of genes from non-crossable species for breeding crops. Cisgenesis involves only genes from the plant itself or from a crossable close relative, and these genes could also be transferred by conventional breeding methods. Crop wild relatives are therefore a valuable source of traits for cisgenesis.

Plant breeders need to understand the various valuation strategies very early in the breeding process if they are to direct long-term selection toward reducing agriculture's negative environmental impacts and achieving greater sustainability while maintaining productivity. Regardless of method, breeding objectives can be broadened to include traits which reduce the environmental footprint of traditional production systems (e.g. nutrient and water use efficiencies that reduce off-farm inputs), to adapt crops to new climates, to host plant resistance to tackle old and emerging pathogen epidemics, or new cultivars for new production systems (e.g. perennial polycultures that mimic the biodiversity of natural systems), albeit with some reduction in rate of gain for the traditional agronomic traits of interest. Interdisciplinary crop improvement strategies accounting for ecological, socio-economic and stakeholder consider-ations will help identify traits leading to plant cultivars using fewer inputs, less land, and less energy, thereby resulting in a more sustainable agricultural ecosystem.

The impact of breeding on crop production is dependent upon the complex relationships involving the farmers, the cultivars available to them, and the developers of those cultivars. Farmers consist of commercial producers with varying size land holdings ranging from moderately small farms to very large ones, and subsistence farmers with small farms often on marginal lands. The subsistence farmers are usually poor. Several types of cultivars are available. The least sophisticated in terms of methods of development are landraces, also known as local cultivars. Modern cultivars consist of development by crossing and selection alone, those developed by crossing and selection with specific important improvements are often obtained from crosses with wild species or by transgenic methods, and F_1 hybrids between desirable inbred lines. The developers of landraces are usually farmers themselves, and are obtained by repeated simple selection procedures of generation after generation. Improved cultivars and hybrids are created either by public sector breeders or seed companies.

Nearly 70% of the world's farmers, from 570 million world exploitations, are small/subsistence and poor farmers. They feed 1,5 billion of the world's population. So they are also a key for biodiversity and for improving the sustainability. For these farmers improved cultivars, hybrids or transgenic seeds tend to be riskier than landraces, since the higher costs associated with seeds and production impose a greater income risk. The lack of capital available denies them the opportunity to invest in production inputs. Small farmers may have lower production costs with landraces because they achieve adequate yields with fewer inputs. In addition, profits from improved hybrid or transgenic cultivars tend to be more variable. Yields are often higher but market prices tend to be inconsistent. For example in India states of Andhra Pradesh and Maharashtra, farmers have been promised higher yields and lower pesticide costs when using *Bt* cotton, thus they acquired loans to afford the costly seeds (Monsanto has control over 95 % of the Indian *Bt* cotton seed market and this near monopoly has resulted in great increased prices). When, in many cases, the farmers found the yields failed to meet their expected result, the consequences were usually very serious and many farmers died by committing suicide over the past 15 years, perhaps due to this reason. This situation of using *Bt* cotton seeds was explained by the absence of irrigation systems combined with specialization in high-cost crops, and played low market prices. Without collateral help these farmers are usually unable to secure a loan from a bank or money lender [43,88]. Rates are often unmanageably high for those able to get a loan, with strict penalties for late payments. Similarly, a lack of education, resources, skill training and support prevent these farmers from using improved cultivars and then to generate a stable income from their production. In addition, governments do not usually regulate the price of crops or even provide market information. Improving market information systems for crops and facilitating farmers' access to credit are then essential components for a strategy to enable poor farmers to grow improved cultivars. A major obstacle to success in crop production using improved cultivars is the shortage of affordable credit. Desperate for cash, subsistence farmers are forced to sell their crops immediately after the harvest to middlemen or their creditors at unfavorable prices. Low cost quality seeds are essential for these poor farmers to improve their life [43].

3.2. Breeding to adapt plants to the environment

3.2.1. Producing more with less

In the coming decades we will need to produce more with less. Fresh water suitable for irrigation is expected to become increasingly scarce and the costs of fertilizer and other agricultural inputs will increase as fossil-fuel costs rise. Nevertheless, continuing gains in production per hectare must be realized to offset the loss of premium agricultural lands (e.g. from urbanization and industrialization), while supplying a growing population. By developing resource efficient plants, plant breeders can continue to improve the sustainability of agricultural ecosystems. Plants requiring fewer off-farm input applications (specifically water, pesticides, nitrogen, phosphorus, and other nutrients) decrease the cost of production, lower fossil energy use, and reduce contamination of water systems, which help to improve public health and stabilize rural economies [95,96].

Although modern plant breeding efforts initially focused on improving uptake of inputs, recent efficiency gains have been made in physiologically increasing yield and biomass production without further increasing inputs. Many crops already have genetic variation in nutrient use efficiency, utilization, and uptake [97-99] and plant breeding will further improve these traits. Intensive agro-ecosystems should emphasize improvements in system productivity, host plant resistance and enhance use-efficiency of inputs such as water and fertilizers.

Water use-efficiency and water productivity are being sought by agricultural researchers worldwide to address water scarcity. Under water scarcity, yields of crops, are a function of how efficiently the crop uses this water for biomass-growth, and the harvest index. Water use efficiency is the ratio of total dry matter accumulation to evapotranspiration and other water losses. An increase in transpiration efficiency or reduction in soil evaporation will increase water use efficiency. Water productivity is the ratio of biomass with economic value produced compared to the amount of water transpired. Both water use efficiency and water productivity may be improved through plant breeding. Farooq et al. [100] discuss the advances in transgenic breeding for drought-prone environments. In their review, they noted the testing of 10 transgenic rice events [unique DNA recombination taking place in one plant cell and thereafter to be used for generating entire transgenic plant(s)] under water scarcity. It seems the transgenic expression of some stress-regulated genes leads to increased water use efficiency.

Agriculture contributes significantly to greenhouse gas emissions. Nitrous oxide and dioxide are potent greenhouse gases released by manure or nitrogen fertilizer, particularly in intensive cropping systems. Nitrous oxide (N_2O), which is a potent greenhouse gas, is generated through the use of manure or nitrogen fertilizer. In many intensive cropping systems nitrogen fertilizer practices lead to high fluxes of N_2O and nitric oxide (NO). Several groups of heterotrophic bacteria use NO_3 as a source of energy by converting it to the gaseous forms N_2, NO, and NO_2 (nitrous dioxide). N_2O is therefore often unavailable for crop uptake or utilization.

Genetic enhancement of crops shows great potential for reducing N_2O emissions from soils into the atmosphere. Some plants possess the capacity to modify nitrification *in situ* because they produce chemicals which inhibit nitrification in soil. This release of chemical compounds from plant roots suppressing soil nitrification has been called biological nitrification inhibition, which seems to vary widely among and within species, and appears to be a widespread phenomenon in some tropical pasture grasses, e.g. *Brachiaria humidicola*. Biological nitrification inhibition may be an interesting target trait of crop genetic engineering for mitigating climate change.

Almost one-fifth of global methane emissions are from enteric fermentation in ruminant animals. Apart from various rumen manipulation and emission control strategies, genetic engineering is a promising tool to reduce these emissions. The amount of methane produced varies substantially across individual animals of the same ruminant species. Efforts are ongoing to develop low methane-emitting ruminants without impacting reproductive capacity and wool and meat quality. A recent study by Shi et al. [101], to understand why some sheep produce less methane than others, deployed high-throughput DNA sequencing and specialized analysis techniques to explore the contents of the rumens of sheep. The study showed that the microbiota present in sheep rumen was solely responsible for the differences among high and low methane emitting sheep. It was further observed that the expression levels of genes involved in methane production varied more substantially across sheep, suggesting differential gene regulation. There is an exciting prospect that low-methane traits can be slowly introduced into sheep.

Crops are bred for nitrogen use efficiency because this trait is a key factor for reducing nitrogen fertilizer pollution, improving yields in nitrogen limited environments, and reducing fertilizer costs. The use of genotypes of same species efficient in absorption and utilization of nitrogen is an important strategy in improving nitrogen use efficiency in sustainable agricultural systems. Crops are being bred for nitrogen use efficiency because this trait will be a key factor for reducing run-off of nitrogen fertilizer into surface waters, as well as, for improving yields in nitrogen limiting environments. There are various genetic engineering activities for improving nitrogen use efficiency in crops [98,102]. The gene Alanine aminotransferase from barley, which catalyzes a reversible transamination reaction in the nitrogen assimilation pathway, seems to be a promising candidate for accomplishing this plant breeding target. Transgenic plants over-expressing this enzyme can increase nitrogen uptake especially at early stages of growth. This gene technology was licensed to a private biotechnology company, and is slated to be commercialized within the next six years [103]. A patent gave this biotechnology company the rights to use the nitrogen use efficiency gene technology in major cereals, as well as, in sugarcane.

Keeping nitrogen in ammonium form will affect how nitrogen remains available for crop uptake and will improve nitrogen recovery, thus reducing losses of nitrogen to streams, groundwater and the atmosphere. There are genes in tropical grasses such as *B. humidicola* and in the wheat wild relative *Leymus racemosus* that inhibit or reduce soil nitrification by releasing inhibitory compounds from roots and suppressing *Nitrosomonas* bacteria [104]. Their value for genetic engineering crops for reducing nitrification needs to be further investigated.

3.2.2. *Adapting to global climate change and for abiotic and biotic stress tolerance*

Extreme weather events are expected to increase in both number and severity in coming years [105]. Climate change impacts agro-ecosystems through changes over the long-term in key variables affecting plant growth (e.g. rising temperatures) and through increasing the variability (frequency and intensity) of weather conditions (rainfall, drought, waterlogging and elevated temperatures). These changes affect both crop productivity and quality. In addition to physically destroying crops, climate change has altered host-pathogen relationships and resulted in increased disease incidence, in insect-pest borne stress in crop plants, and in invasive pests which feed and damage them.

There are two ways to adapt crops to new environments: developing new crops (long-term endeavor starting with domestication) and introducing target traits into existing crops through plant breeding, which includes genetic engineering. However, the job of crop improvement is becoming increasingly difficult. Cultivars which are not only high yielding but are also efficient in use of inputs are needed, tailored to ever more stringent market demands, able to maintain stability under increasing climate variability, and potentially contribute to climate mitigation. These multi-trait demands for new cultivars provide significant challenges for crop breeders, and standard selection approaches struggle under such complexity. To maintain productivity in the face of increased climatic variability, both the population and the plant cultivars will need to be continually developed to withstand "new" climate extremes and the stresses which these will entail [106].

Many breeding programs are already developing plants which tolerate extreme weather conditions, including drought, heat, and frost [107,108]. Plant breeders are also beginning to address expected changes due to increased climate variability, by increasing genetic diversity sources and by adjusting selection and testing procedures [109].

More frequent weather extremes will likely affect the existing ranges of not only agronomic cultivars but also local native plant species [110]. Because some genetic variation useful for climate change adaptation will be found only in wild plant relatives of cultivated crops, preserving genetic biodiversity is essential in order for breeders to select plants that will be well-suited for future environmental conditions [111].

Global climate change notwithstanding, additional stress tolerances in crop species are needed to maintain productivity and survival. In the near future, tolerance to various soil conditions including acidic, aluminum-rich soils (particularly in the tropics) and saline soils (especially those resulting from irrigation), will be increasingly important for production on marginal agricultural lands or as the salt content of irrigated lands increases [112]. Bhatnagar-Mathur et al. [113] suggested that genetic engineering could accelerate plant breeding to adapt crops to stressful environments. They further underline that engineering the regulatory machinery involving transcription factors (TF; a protein binding specific DNA sequences and thereby governing the flow of genetic information from DNA to messenger RNA) provides the means to control the expression of many stress-responsive genes. There are various target traits for adapting crops, through genetic engineering, to high CO_2 and high O_3 environments of the changing climate [114]. Ortiz [115], Jewell et al. [116], and Dwivedi et al. [117,118] provide the most recent overviews on research advances in genetic engineering for improved adaptation to drought, salinity or extreme temperatures in crops. The most cited include TF, and genes involved in: i) signal sensing, perception, and transduction; ii) stress-responsive mechanisms for adaptation; and iii) abscisic acid biosynthesis for enhanced adaptation to drought. Transporter, detoxifying and signal transduction genes as well as TF are cited for tolerance to salinity. Genes related to reactive oxygen species, membrane and chaperoning modifications, late abundance embryogenesis proteins, osmoprotectants/compatible solutes and TF are pursued in crop genetic engineering for temperature extremes.

Transgenic crops provide the means to adapt crops to climate change, particularly in terms of drought and salinity. Duration and intensity of drought has increased in recent years, consistent with expected changes of the hydrologic cycle under global warming. Drought dramatically reduces crop yields. Genetic engineering may be one of the biotechnology tools for developing crop cultivars with enhanced adaptation to drought [119]. It should be seen as complementary to conventional plant breeding rather than as an alternative to it. The function of a TF such as the *Dehydration-Responsive Element Binding* (*DREB*) gene in water stress-responsive gene expression has been extensively investigated [120]. The main research goal was to gain a deep understanding of TF in developing transgenic crops targeting drought-prone environments [121]. For example, the *DREB1A* gene was placed under the control of a stress-inducible promoter from the *rd29A* gene and inserted via biolistic transformation into wheat bread [122]. Plants expressing this transgene demonstrated significant adaptation to water stress when compared to controls under experimental greenhouse conditions as manifested by a 10-day delay in wilting when water was held. Saint Pierre et al. [123] indicated, however, that these transgenic lines did not generally out-yield the controls under water deficit in confined field trials. Nonetheless, they were able to identify wheat lines combining acceptable or high yield under enough irrigation which also showed stable performance across the water deficit treatments used in their experiments; i.e., severe stress, stress starting at anthesis, and terminal stress.

Soils affected by salinity are found in more than 100 countries, and about 1/5 of irrigated agriculture is adversely affected by soil salinity. Therefore, breeding salt-tolerant crops should be a priority because salinity will most likely increase under climate change. Mumms [124] lists some candidate genes for salinity tolerance, indicating the putative functions of these genes in the specific tissues in which they may operate. Genes involved in tolerance to salinity in plants, limit the rate of salt uptake from the soil and the transport of salt throughout the plant, adjust the ionic and osmotic balance of cells in roots and shoots, and regulate leaf development and the onset of plant senescence. The most promising genes for the genetic engineering of salinity tolerance in crops, as noted by Chinnusamy et al. [125], are those related to ion transporters and their regulators, as well as the C-repeat-binding factor. The recent genome sequencing of *Thellungiella salsuginea*, a close relative of *Arabidopsis* thriving in salty soils, will provide more resources and evidence about the nature of defense mechanisms constituting the genetic basis underlying salt tolerance in plants [126].

In the quest for breeding transgenic rice and tomato, advances showing salt tolerance have occurred. Plett et al. [127] were able to show an improved salinity tolerance in rice by targeting changes in mineral transport. They initially observed that cell type-specific expression of *AtHKT1* (a sodium transporter) improved sodium (Na^+) exclusion and salinity tolerance in *Arabidopsis*. Further research explored the GAL4-GFP enhancer trap (transgenic construction inserted in a chromosome and used for identifying tissue-specific enhancers in the genome) to drive expression of *AtHKT1* in the root cortex in transgenic rice plants. The transgenic rice plants had a higher fresh weight under salinity stress due to a lower concentration of Na^+ in the shoots. They also noted that root-to-shoot transport of $^{22}Na^+$ decreased and was correlated with an up- regulation of *OsHKT1*, the native transporter responsible for Na+ retrieval from the transpiration stream. Moghaieb et al. [128] bred transgenic tomato plants producing ectoine (a common compatible solute in bacteria living in high salt concentrations). Ectoine synthesis was promoted in the roots of transgenic tomato plants under saline conditions, which led to increased concentration of photosynthesis in improving water uptake. Likewise, the photo-synthetic rate of ectoine-transgenic tomato plants increased through enhancing cell membrane stability in oxidative conditions under salt stress.

Transgenic crops can also contribute to climate change mitigation efforts by reducing input use intensity [129]. The integration of genetic engineering with conventional plant breeding, within an interdisciplinary approach, will likely accelerate the development and adoption of crop cultivars with enhanced adaptation to climate change related stresses [130]. Global warming will reduce yields in many crops about 6% and 5% average yield loss per 1°C in C_3 and C_4 crops, respectively, whose optimum temperature ranges are 15–20°C and 25–30°C [131]. The extent of yield loss depends on crop, cultivar, planting date, agronomy and growing area. For instance, an increase of 1°C in the night time maximum temperature translates into a 10% decrease in grain yield of rice, whereas a rise of 1°C above 25°C shortens the reproductive phase and the grain-filling duration in wheat by at least 5%, thereby reducing grain yield proportionally. Heat stress will exacerbate climate change impacts in the tropics, while it may put agriculture at risk in high latitudes where heat-sensitive cultivars are grown today. Hence, new cultivars must be bred to address heat stress. Ainsworth and Ort [132] suggested giving priority to traits improving photosynthesis for adapting to heat stress. However, plants have various mechanisms to cope with high temperatures, e.g. by maintaining membrane stability, or by ion transporters, proteins, osmoprotectants, antioxidants, and other factors involved in signaling cascades and transcriptional control [133,134]. Furthermore, Gao et al. [135] noted that *bZIP28* gene (a gene encoding a membrane-tethered TF) up-regulated in response to heat in *Arabidopsis*. Some of these genes can be used in crop genetic engineering to enhance plant adaptation to heat stress. For example, some stress-associated genes such as ROB5, a stress inducible gene isolated from bromegrass, enhanced performance of transgenic canola and potato at high temperatures [136]. Likewise, Katiyar-Agarwal et al. [137] introduced *hsp101* gene (a heat shock protein gene from *Arabidopsis*) in basmati rice. This transgenic rice had a better growth in the recovery phase after suffering heat stress.

Globalization has, among other consequences, led to the rapid spread of plant disease and invasive pests. Being immobile, plants are unable to escape pathogens causing plant disease and pests which feed and damage them. Plant disease is mainly caused by fungi, bacteria, viruses, and nematodes. Approximately 70,000 species of pests exist in the world, but of these, only 10% are considered serious [138]. Synthetic pesticides have been applied to crops since 1945 and have been highly successful in reducing crop losses to some pest insects, plant pathogens, weeds and in increasing crop yields [138]. One estimate suggests that without pesticides, crop losses to pests might increase by 30%. Despite pesticide use, insects, pathogens and weeds continue to exact a heavy toll on world crop production, approaching 40% [138,139]. Pre-harvest losses are globally estimated at 15% for insect pests, 13% for damage by pathogens, and about 12% for weeds [138]. Developing resistant cultivars reduces the need for expensive and environmentally damaging pesticides to be applied. For example, a recent outbreak of *Xanthomonas campestris* pv. musacearum led to the devastating *Xanthomonas* wilt of banana in the Great Lakes Region of Africa, thereby threatening the food security and income of millions of East and Central African people who depend on this crop. Transgenic banana plants with the hypersensitivity response-assisting protein (*Hrap*) gene from sweet pepper did not show any infection symptoms after artificial inoculation of potted plants with *Xanthomonas* wilt in the screen house [140]. Selected transgenic banana plants with putative host plant resistance to *Xanthomonas* wilt are ongoing confined field-testing in East Africa, where elevated temperatures, due to the changing climate, will likely favor banana production.

Weather influences how pathogens and pests affect and interact with crops and their host plant resistance, and thus climate change can also have wide-ranging impacts on pests and diseases [118]. Late blight, which is caused by *Phytophthora infestans*, ranks as the most damaging potato pest. Late blight accounts for 20% of potato harvest failures worldwide, translating into 14 million tonnes valued at 7.6 billion US dollars. Global warming will increase late blight spread, e.g. expanding its range above 3,000 meters in the Andes [141]. Chemical control may lead to more aggressive strains of the pathogen and chemical control is often regarded as being environmentally damaging. Cisgenic potato cultivars with late blight resistance are becoming available and will impact growers, consumers and the environment favorably [142]. Related wild *Solanum* species can be a source of alleles to enhance host plant late blight resistance in potato. For example, *S. bulbocastum* (a wild relative with high resistance to late blight from Mexico) was used to breed the cultivar "Fortuna" using genetic engineering. Cisgenesis allows inserting several host plant resistance genes from wild crop species in one step without linkage drag (reduction in cultivar fitness).

3.3. Breeding plants to improve the environment

In general, plants are bred for their most obvious end products, including grain, fiber, sugar, biomass yield, fruit quality, or ornamental qualities. However, plants deployed across the landscape in agricultural or forestry settings affect the environment in measurable ways. Perennial crops have environmentally beneficial properties not present in annual crops, such as helping to prevent erosion in agricultural systems, providing wildlife habitat, and acting as sinks for carbon and nutrients. Traditionally, perennial crops have not been a major focus of breeding programs because they generally take more time and scientific knowledge to improve, and therefore, products such as new cultivars are often not produced within the timeframe of funding cycles. Current tree breeding programs are developing elms (*Ulmus* spp), chestnuts (*Castanea dentata*), hemlocks (*Tsuga* spp), and other species which are resistant to introduced diseases and insects [143,144]. As compared with natural selection, artificial selection via plant breeding has overcome these stresses more effectively by rapidly incorporating diverse exotic genetic sources of resistance, hybridizing to include multiple, different genetic resistances into the same plant, and making use of off-season locations or artificial conditions to shorten generation cycles. A more complex example which may be feasible in the future is tree breeding for larger and improved root systems to decrease soil erosion, sequester carbon, and improve soil quality by increasing soil organic matter.

New crop cultivars developed by plant breeders must help improve soil health, reduce soil erosion, prevent nutrient and chemical runoff, and maintain biodiversity. The goal to breed projects for forages, which include several species, is to produce a high yield of leaf and stem biomass, as opposed to grain, for ruminant animals. In the tropics many forages are perennial, providing year-round erosion control, improving water infiltration as compared with that, from annual cropping systems, and in some cases, sequestering carbon. The forage breeding program at the University of Georgia (UG) has developed cultivars in several species and has been proactive in developing agreements with private-sector commercial partners to oversee seed production and marketing of new cultivars. Among the cultivars developed at UG is "Jesup MaxQ" tall fescue, a cultivar carrying a non-toxic endophytic fungus that was both highly persistent under grazing and greatly improved animal weight gain and feed efficiency over standard cultivars. In addition, this program developed the first true dual purpose, grazing and hay, alfalfa cultivar "Alfagraze", followed by several further improved alfalfa cultivars like "Buldog 805" which persist through summer under cattle grazing [145].

Cover crops are annual species planted in rotation with crops to specifically improve soil conditions and to control weeds, soil-borne diseases, and pests [146-148]. Continuous cover crops can reduce on-farm erosion, nutrient leaching, and grain losses due to pest attacks and build soil organic matter as well as improve the water balance, leading to higher yields

[149,150]. For instance, in Kenya, Kaumbutho and Kienzle [151] showed in two case studies that maize yield increased from 1.2 to 1.8–2.0 t/ha with the use of mucuna legume as cover crop, and without application of nitrogen fertilizer. Besides farmers who adopted mucuna legume as cover crop benefited from higher yields of maize with less labor input for weeding.

Many current perennial and cover crop cultivars are essentially wild species bred from germplasm collections and developed to increase success in managed agro-ecosystems. As compared with non-native vegetation, plant species native to a particular region are generally thought to survive on less water, use fewer nutrients, require minimal pesticide applications, and be non-invasive; however, counter examples for both native and non-native species are plentiful [152]. As potentially valuable species are identified, breeding to improve them for traits of consumer importance will be needed to broaden available biodiversity in cultivated landscapes. With a changing climate, species considered critical to the landscape may require human-assisted hybridization with distant relatives to better ensure survival from threats posed by novel pests or diseases.

Alternative crops are also being bred for new uses, such as removing toxic chemicals and excess nutrients and improving degraded soils, including mine spoils [153]. Phytoremediation is a biotechnology to clean the contaminated sites of toxic elements (e.g. Cd, Cu, Zn, As, Se, Fe) via plant breeding, plant extracting, and plant volatilizing [154]. The last few years have seen a steady expansion in the list of hyper-accumulator species, which could be valuable plant resources for phytoremediation. For example an ecotype of the Zn/Cd hyper-accumulator *Thlaspi caerulescens* from southern France was able to phytoextract Cd efficiently in field trials through the different seasons with good growth of biomass [155,156]. The Chinese brake fern *Pteris vittata* has a strong ability to hyper-accumulate arsenic (As) and shows promising potential for phytoextraction of and from contaminated soils under field conditions [157,158].

A major goal of harmonizing agriculture with the environment is to "tailor" crops to individual landscapes. Plant breeding has always maximized production by selecting for adaptation in the target environments of interest, using local environmental forces for plant selection [131]. By selecting breeding germplasm growing under local environmental conditions, individual cultivars can be optimized for small regional areas of production that fit prevailing environmental and weather patterns. Likewise, plants could be tailored to provide specific ecosystem services to local environments, to address local needs. One cost-effective way to achieve this is through participatory plant breeding, which involves local farmers in the breeding process.

Alternative crop rotations, planting densities, and tillage systems may make production more environmentally benign but will require altering breeding targets and an understanding that systems biology is complex and rarely has simple solutions. For example, no-tillage systems used for soil conservation can lead to colder soils in spring and change the prevalence and onset of various soilborne diseases, thus requiring the addition of specific disease resistances in the breeding objectives [159]. Breeders must select from conditions prevailing under new management practices to ensure cultivars will be optimally productive.

4. Conservation and use of biodiversity – opportunities for cooperation and new partnerships

Plant genetic resources for food and agriculture are the quintessential global public good. No nation is self-reliant. A viable market for their conservation and trade does not exist. The conservation of plant genetic resources is a prerequisite for addressing climate change, as well as water and energy constraints, which will grow in importance in the next decades. The Svalbard Global Seed Vault is an International Treaty which establishes a multilateral System to facilitate access and benefit sharing of plant genetic resources. The Treaty has an insurance policy and provides legal framework for a cooperative and global approach to manage this essential resource. The Svalbard Global Seed Vault has a mechanism for ensuring the permanent conservation of unique crop biodiversity, the Global Crop Diversity Trust, which is structured as an endowment fund [45].

Plant breeding is vital to increase the genetic yield potential of all crops. As menthioned a result of the Green Revolution was the increase of global productivity of the main food staples. Such achievements ensued from crop genetic enhancement partnerships. These partnerships include national agricultural research institutes and international agricultural research centers. For many decades the global wheat yield increased due to an effective International Wheat Improvement Network (IWIN) officially founded as an international organization in 1966 [160]. This wheat network deployed cutting-edge science alongside practical multi-disciplinary applications, resulting in the development of genetically enhanced wheat germplasm, which has improved food security and the livelihoods of farmers in the developing world [161]. The spring wheat germplasm bred in Mexico under the leadership of Nobel Peace Laureate Norman Borlaug was further used for launching the Green Revolution in India, Pakistan and Turkey [162]. The network was broadened during the 1970s to include Brazil, China and other major developing country wheat producers. It resulted in wheat cultivars with broader host plant resistance (especially to rusts), better adaptation to marginal environments, and tolerance to acid soils. Nowadays IWIN, an international "alliance", operates field evaluation trials in more than 250 locations, in roughly 100 countries it tests improved breeding lines of wheat in different environments. The number of wheat cultivars released annually in the developing world doubled to more than 100 cultivars by early 1990s due to this networking and the strengthening of national capacity [163]. The widespread adoption of newly bred wheat cultivars, especially in South Asia and Latin America, due to yield increases, led to 50% average annual rates of investment returns [164]. The urban poor also benefited significantly because grain harvest increases drove wheat prices down. Every year, nursery sets and trials are sent to various researchers worldwide, who share their data from these trials to catalogue and analyze. The returned data are used to identify parents for subsequent crosses and to incorporate new genetic variability into advanced wheat lines that are consequently able to cope with the dynamics of abiotic and biotic stresses affecting wheat farming systems. The full pedigree and selection histories are known and phenotypic data cover yield, agronomic, pathological and quality data [161].

The International Network for Genetic Evaluation of Rice (INGER) is one more example of world cooperation. It was established in 1975 as a consortium of national agricultural research systems of rice-growing countries and Centers of today's CGIAR Consortium. INGER was initially founded as an International Rice Testing Program, but soon became an integral component of world national rice breeding program. INGER partners can share rice breeding lines. Every year partners provide about 1000 genetically diverse breeding lines, which have been grown in about 600 experiment stations from 80 countries. This network facilitated the release of 667 cultivars worldwide, which translated into 1.5 billion US dollars of economic benefits. It was estimated that ending INGER could lead to a reduction of 20 rice cultivars per year and to an economic loss of 1.9 billion US dollars [165]. Further analysis by Jackson and Huggan [166] has shown how genetic conservation of landraces can lead to significant gains in rice breeding.

Two other examples of cooperation and partnership are the Latin American Maize Project (LAMP) and the Germplasm Enhancement of Maize (GEM). The LAMP was established as a partnership between Latin America and the United States to assess national germplasm and facilitate the exchange of maize genetic resources across the American continent [167]. The United States Department of Agriculture, the participating national agricultural research systems and a multinational seed corporation provided the funding. The aim of LAMP was to obtain information about the performance of maize germplasm and to share it with plant breeders for developing genetically enhanced open pollinated and hybrid cultivars. The maize germplasm was tested for agronomic characteristics from sea level to 3300 m, and from 41°N to 34°S across 32 locations in the first stage and in 64 locations (two per region) in the second stage. These locations were clustered according to five homologous areas: lowland tropics, temperate and three altitudes.

There were a total five LAMP breeding stages [167]. In the first stage, 14,847 maize accessions belonging to a region were planted for evaluation in trials using a randomized complete block design with two replications of 10m^2 plots at a single location, which was environmentally similar to that from where these landraces were originally collected. The next step included the assessment of the upper quintile (20%) of those accessions evaluated for agronomic performance in the previous stage. These accessions were planted in two locations with two replications, and the upper 5% were further selected according to their performance. These best selected accessions of each country were interchanged among regions belonging to the same homologous area in the third stage. They were tested in two locations with two replications in each region. The selected maize accessions from the same homologous area were mated with the best tested accession of the region in an isolated field within each region. In the fourth stage, combining ability tests of 268 selected maize accessions were carried out with a local tester using two replications at two locations within each region. The elite maize germplasm was integrated into breeding programs in the fifth stage, which was the last. The best cross combinations and heterotic pools were also determined by LAMP. Maize breeders obtained access to the most promising accessions identified by LAMP to widen the crop genetic base. A LAMP core subset has been made available for encouraging further use in broadening of maize genetic diversity [168].

The GEM was set up to introgress useful genetic diversity from Latin American maize landraces and other tropical maize donor sources (lines and hybrids) into United States' maize germplasm, to broaden the genetic base of the "corn-belt" hybrids [169,170]. GEM owes its existence to LAMP because it has used the Latin American landrace maize accessions selected by LAMP in crosses with elite temperate maize lines from the private seed companies in North America [167]. GEM used a pedigree breeding system to develop S_3 lines. The GEM breeders arranged their crosses into non-Stiff Stalk and Stiff Stalk heterotic groups [171].

LAMP provided the first step through the sharing of information needed to select gene bank maize accessions for further germplasm enhancement. GEM completed the process by returning to genetically enhanced breeding materials derived from gene bank accessions. This improved germplasm can be further used in maize breeding in the United States and elsewhere. LAMP and GEM are very nice examples of international and national public-private partnerships in crop germplasm enhancement.

Agricultural plant breeding is a typical commodity- or species-oriented and solves problems within a species, rather than making breeding choices based on system wide needs. For example, maize breeders currently maximize the area in which maize can be grown, and maximize the amount of maize produced throughout that area. If environmental harmony is to be a key breeding objective, then a change in agricultural thinking to appropriately value whole cropping systems will be required. Achieving these goals will require collaboration among the private, public, and non-profit sectors, and with society as a whole. Programs within the private sector excel at breeding major, profitable crops, and have economies of scale to increase the efficiency of production and ultimately provide farmers with seed. As a valuable complement to commercial breeding programs, public and non-profit breeding programs may focus on developing alternative crops, breeding for small target regions, tackling long-term and high-risk problems, evaluating diverse genetic resources, and, importantly, conducting basic research on breeding methodology to enhance efficiency. Only publicly funded breeding programs, and in particular those based at universities, can provide the necessary education and training in plant breeding and in specialized fields such as ecology. Without trained students from public programs, private commercial breeding programs suffer from an erosion of intellectual capital. Conversely, without the private sector to commercialize public-sector-derived products, beneficial traits and new cultivars cannot easily and quickly be put in the hands of farmers, as has been seen in developing countries without a developed seed industry [172]. As stated, seed production is high technology and a cost intensive venture and only well organized seed companies with good scientific manpower and well equipped research facilities can afford seed production.

Although due to globalization, most breeding research and cultivar development in the world is presently conducted and funded in the private sector, mainly by huge multinational seed companies. Public breeders, cultivar development activities and research are disappearing worldwide. In general, this means there are fewer decision-making centers for breeding and cultivar development. This has also resulted in the focus on relatively few major crops produced worldwide, to the detriment of all the other cultivated crops. It is imperative that national governments and policymakers, as part of a social duty, invest in breeding research and cultivar development of traditional open-pollinated cultivars and in the minor crops. More investments in this area will mean less expensive seed for growers to choose from, and an increased preservation of crop biodiversity. To accomplish these goals new approaches may be required to crop breeding research and development by both the public and private sector. Until recently, breeding research and development which targets small-scale and poor farmers has largely been undertaken by public sector institutions and national agricultural research institutes. However, the capacity to undertake the work was mainly dependent on national or international funding and expertise. The work has been limited by the capacity of these institutions to pay for it. As a result, crop breeding advancement has varied enormously among countries and even within regions in developed and still developing countries. In the area of plant breeding, the process to produce improved cultivars is slow, and it requires long-term sustained commitment that may not fit the continuing changes in the national and international politics to fund research. The application of biotechnology promises acceleration in some aspects of plant breeding, but the adoption of more advanced technology raises the cost of research significantly at a time when investment funding has diminished. Public plant breeding remains a key component of crop breeding research systems worldwide, especially in developing countries. However, the increasing presence of private sector breeding and a decrease in national and international support makes it difficult for the public sector to continue operating in the traditional manner. Declining funding for public crop breeding coupled with the rapid increase of crop production and an urbanizing population has created a difficult situation. Public sector breeding must be strengthened. More public sector crop breeders are needed worldwide to select and to produce non-hybrid cultivars of the minor crops. Breeding of major crops and other minor crops must continue as a viable endeavor. This will benefit small farmers, and will safeguard biodiversity and food security in developing countries.

While the maintenance of vigorous public sector breeding programs in areas where private companies are not interested in providing low cost cultivars is highly desirable, an additional approach to maximize crop and agricultural research input would be the development of global programs with public-private partnerships. The public sector may support portions of crop and agricultural R&D, unattractive to the private sector, and feed improved breeding lines and systems to the private sector for exploitation in regions where the private sector is active, and nurture private sector development in regions where it is lacking. In recent years, private plant breeding programs have increased in number and size. Financial investment also increased, as well as interest in intellectual property protection. The spirit of original attempts to protect plant breeders' rights was that granting a certificate of protection should not inhibit the flow of information and products through continued research by the entire plant breeding community [106,107]. In a classic sense, the patent is a defensive tool to prevent competitors from reaping benefits which rightfully belong to the inventor. In the modern context, it is an offensive weapon, to stifle competition, prevent further innovation by others and maximize income [106,108]. The United States utility patent, it is a way to slow down the flow of progress in plant breeding research, unless the research is within the company holding the patent. While obviously benefiting that company, it is a big step backwards for the plant breeding community and by far, for agriculture itself. The intellectual property protection must encourage research and free flow of materials and information [106,108]. Protection should be for the cultivar only. There should be no constraint against other breeders using that cultivar in further research, including further breeding. Another breeder should be free to use the protected cultivar in a cross, followed by further development through pedigree breeding. Another breeder should also be free to transfer genes controlling economic traits into the protected cultivar by the backcross method or by genetic transformation procedures [106,107].

5. Conclusions

The growing demand for food in the next decades poses major challenges to humanity. We have to safeguard both arable land for future agricultural food production, and protect genetic biodiversity to safeguard ecosystem resilience. Besides we need to produce more food with less inputs.

Plant breeding is the science of improving plants to further improve the human condition. Plant breeding has played a vital role in the successful development of modern agriculture via "new" cultivars. Plant breeders are continually improving the ability of cultivars to withstand various environmental conditions. By reducing the impact of agriculture on the environment while maintaining sufficient production will require the development of new cultivars.

Climate change is altering the availability of resources and the conditions crucial to plant performance. Plants respond to these changes through environmentally induced shift in phenotype. Understanding these responses is essential to predict and manage the effects of climate change on crop plants.

In the foreseeable future and an increase in population will need significant production. Breeding and modern agricultural technologies can increase yield on existing agricultural land. As a result, they can make a significant contribution to biodiversity conservation by limiting the need to expand agricultural land and by allowing nature to be maintained for conservation purposes and harmony between agriculture and the environment.

There is still a debate among researchers on the best strategy to keep pace with global population growth and increasing food demand. One strategy focuses on agricultural biodiversity, while another strategy favors the use of transgenic crops. There are short research funds for agro-biodiversity solutions in comparison with funding for research in genetic modification of crops. Favoring biodiversity does not exclude any future biotechnological contributions, but favoring biotechnology threatens future biodiversity resources. The future breeding programs should encompass not only knowledge of techniques but also conservation of genetic resources of existing crops, breeds, and wild relatives, to provide the genes necessary to cope with changes in agricultural production. Therefore, agro-biodiversity should be a central element of future sustainable agricultural development [173,174]. The concept of sustainability rests on the principle that the present needs must be addressed without compromising the ability of future generations to meet their own needs [175]. Sustainable agriculture is an alternative to solve future fundamental and applied issues related to food production in an ecological way [176].

Farmers in developing countries, especially small farmers, have problems specific to their cultural, economic and environmental conditions, such as limited purchasing power to access improved cultivars and proprietary technologies [43]. These farmers have an important role in conserving and using crop biodiversity. The future of the world food security depends on stored crop genes as well as on farmers who use and maintain crop genetic diversity on a daily basis. In the long run, the conservation of plant genetic diversity depends not only on a small number of institutional plant breeders and seed banks, but also on the vast number of farmers who select, improve, and use crop diversity, especially in marginal farming environments. Their extensive farming systems using landraces or open-pollinated cultivars increase sustainability and less impact from stresses caused by drought, insect and diseases, due to long-term *in situ* selection of these crops cultivated as opposed to the fertilizer, herbicide, and pesticide demands in an intensive crop based system with improved, hybrid, or transgenic cultivars. That is why we should also be alerted and particularly alarmed by the current trend to exclusively use improved, hybrid, and transgenic crop cultivars. Farmers do not just save seeds; they are plant breeders who constantly adapt their crops to specific farming conditions and needs. This genetic biodiversity is the key to maintain and improve the world's food security, and agriculture sustainability [51].

The introduction of genetically modified technology has been hailed as a gene revolution similar to the "Green Revolution" of the 1960s [41,177]. The "Green Revolution" had an explicit strategy for technology development and diffusion, targeting farmers in developing countries, in which improved germplasm was made freely available as a public good, a particular success in Asia. In contrast to the "Green Revolution", the push for genetically modified crops is based largely on private agricultural research, with cultivars provided to farmers on market terms [177]. To date efforts on genetically modified crops have been focused on crops considered to be profitable enough by large plant breeding companies, not on solutions to problems confronted by the world's small farmers. Existing biodiversity in combination with plant breeding has much more to offer the many world's farmers, while genetically modified crops have more to offer the large-scale farms and agro-industry, and this explains why they have received so much research funding. Genetically modified crops and their creation may attract investment in agriculture, but it can also concentrate ownership of agricultural resources. There is particular concern that present intellectual property rights instruments, including genetically modified organisms, will inhibit sowing of own seeds, seed exchange, and sale [178]. And in developing countries, patents may drive up costs and restrict experimentation by the public researcher or individual farmer.

Transgenic crops can continue to decrease pressure on biodiversity as global agricultural systems expand to feed a growing world population. Continued yield improvements in crops such as rice and wheat are expected with insect resistant and herbicide tolerant traits that are already commercialized in other transgenic crops. Although the potential of currently commercialized genetically modified crops to increase yields, decrease pesticide use, and facilitate the adoption of conservation tillage has yet to be realized in some many countries that have not yet approved these technologies for commercialization. Technologies such as drought tolerance and salinity tolerance would alleviate the pressure in arable land by enabling crop production on sub-optimal soils. Drought tolerance technology is supposed to be commercialized within less than three years. Nitrogen use efficient technology is also under development, which can reduce run-off of nitrogen fertilizer into surface waters. This technology is supposed to be commercialized within the next six years.

One of the major arguments for genetic modified technology is that new cultivars can be developed more quickly than in traditional plant breeding [111,116,179]. But like new cultivars derived from conventional breeding methods, transgenic cultivars developed under laboratory conditions have to be tested under field conditions and this means several years of field trials to ensure that the inserted traits will actually become expressed and have the desired effects in local environments. So currently there is little difference in the speed with which either method (transgenic or conventional) will result in the release of new cultivars.

The knowledge gained from basic plant research will underpin future crop improvements, but effective mechanisms for the rapid and effective translation of research discoveries into public good agriculture remain to be developed. Maximum benefit will be derived if robust plant breeding and crop management programs have ready access to all the modern crop biotechnological techniques, both transgenic and non-transgenic, to address food security issues. This will require additional investments in capacity building for research and development, in developing countries. Technology implementation alone is not sufficient to address such complex questions as food security. Biotechnologies will make new options available but are not a global solution. We must ensure that society will continue to benefit from the vital contribution that plant breeding offers, using both conventional and biotechnological tools. Genetic engineering has the potential to address some of the most challenging biotic constraints faced by farmers, which are not easily addressed through conventional plant breeding alone. Besides other promising traits seems to be host plant resistance to insects and pathogens. However, transgenic cultivars will have one or a few exogenous genes whereas the background genotype will still be the product of non-transgenic (or conventional) crop breeding. One should follow a pragmatic approach when deciding whether to engage in transgenic plant breeding. Biotechnology products will be successful if clear advantages and safety are demonstrated to both farmers and consumers.

There is a need of investment in research breeding and cultivar development in traditionally open-pollinated cultivars and in the minor crops. More investments in this area will mean cheaper cultivars for growers to choose from and more preservation of crop biodiversity. In recent years, private plant breeding programs have increased in number and size. Financial investment also increased, as well as interest in intellectual property protection. Protective measures, especially patenting, must be moderated to eliminate coverage so broad that it stifles innovation. The intellectual property protection laws for plants must be made less restrictive to encourage research and free flow of materials and information. Public sector breeding must remain vigorous, especially in areas where the private sector does not function. This will often require benevolent public/private partnerships as well as government support. Intellectual property rights laws for plants must be made less restrictive to encourage freer flow of materials. Active and positive connections between the private and public breeding sectors and large-scale gene banks are required to avoid a possible conflict involving breeders' rights, gene preservation and erosion. Partnerships between policy makers with public and private plant breeders will be essential to address future challenges. Many current breeding efforts remain under-funded and disorganized. There is a great need for a more focused, coordinated approach to efficiently utilize funding, share expertise, and continue progress in technologies and programs.

Author details

João Carlos da Silva Dias

Address all correspondence to: mirjsd@gmail.com

University of Lisbon, Instituto Superior de Agronomia, Tapada da Ajuda, Lisbon, Portugal

References

[1] Jaggard, KW, Qi A, Ober, ES. Possible changes to arable crop yields by 2050. Phiols. Trans. R. Soc. Biol. Sci. 2010; 365:2835-2851.

[2] FAO. The state of the world population report. By choice, not by chance: family planning, human rights and development. United Nations Population Fund, New York; 2012.

[3] Dias JS. Guiding strategies for breeding vegetable cultivars. Agricultural Sciences 2014; 5(1):9-32.

[4] Dias JS, Ryder EJ. World vegetable industry: production, breeding, trends. Horticultural Reviews 2011; 38:299-356.

[5] Dias JS. 1. Vegetable breeding for nutritional quality and health benefits. In: Carbone, K (Ed.). Cultivar: chemical properties, antioxidant activities and health benefits. Nova Science Publishers, Inc., Hauppauge, New York; 2012. Pp.1-81.

[6] Tilman D, Balzer C, Hill J, Befort BL. Global food demand and the sustainable intensification of agriculture Proc. Natl. Acad. Sci. USA 2011; 108:20260-20264.

[7] Alexandratos N, Bruinsma J 2012 World agriculture towards 2030/2050: the 2012 revision No. 12-03. Food and Agriculture Organisation (FAO), Rome; 2012.

[8] FAO. How to feed the world in 2050. FAO, Rome; 2012.

[9] Delgado CL. Rising consumption of meat and milk in developing countries has created a new food revolution J. Nutr. 2003; 133:3907S-3910S.

[10] Kastner T, Rivas MJI, Koch W, Nonhebel S. Global changes in diets and the conse-quences for land requirements for food. Proc. Natl. Acad. Sci. USA 2012; 109:6868-6872.

[11] Delgado CL. Livestock to 2020: the next food revolution. Food, Agriculture, and the Environment Discussion Paper No. 28. International Food Policy Research Institute, Washington, DC; 1999.

[12] The Royal Society. Reaping the benefits: science and the sustainable intensification of global agriculture. The Royal Society Policy Document 11/09. The Royal Society, Lon-don; 2009.

[13] Gerbens-Leenes P, Nonhebel S. Consumption patterns and their effects on land re-quired for food. Ecol. Econ. 2002; 42:185-199.

[14] Wirsenius S, Azar C, Berndes G. How much land is needed for global food produc-tion under scenarios of dietary changes and livestock productivity increases in 2030? Agric. Syst. 2010; 103:621-638.

[15] Pimentel D, Pimentel M. Sustainability of meat-based and plant-based diets and the environment Am. J. Clin. Nutr. 2003; 78:660S-663S.

[16] Steinfeld H, Gerber P, Wassenaar T, Castel V, Rosales M, de Haan C. Livestock's long shadow. Environmental issues and options. FAO, Rome; 2006.

[17] Mekonnen MM, Hoekstra AY. A global assessment of the water footprint of farm an-imal products. Ecosystems 2012; 15:401-415.

[18] FAO. FAO statistical yearbook - Land use. FAOSTAT, FAO, Rome; 2013. PA4.

[19] Kucharik CJ, Serbin SP. Impact of recent climate change on Wisconsin corn and soy-bean yield trends. Environ. Res. Lett. 2008; 3:034003 (10 pp).

[20] Battisti DS, Naylor RL. Historical warnings of future food insecurity with unprece-dented seasonal heat. Science 2009; 323:240-244.

[21] Schlenker W, Lobell DW. Robust negative impacts of climate change on African agri-culture. Environ. Res. Lett. 2010; 5:014010 (8 pp).

[22] Roudier P, Sultan B, Quirion P, Berg A. The impact of future climate change on west African crop yields: what does the recent literature say? Glob. Environ. Change 2011; 21:1073-1083.

[23] Lobell DB, Schlenker W, Costa-Roberts J. Climate trends and global crop production since 1980. Sci. Express; 2011. http://dx.doi.org/10.1126/science.1204531.

[24] Lobell DB, Bänziger M, Magorokosho C, Vivek B. Nonlinear heat effects on African maize as evidenced by historical yield trials. Nat. Clim. Change 2011; 1:42-45.

[25] Schlenker W, Roberts MJ. Nonlinear temperature effects indicate severe damages to US crop yields under climate change. Proc. Natl. Acad. Sci. USA 2009; 106:15594-15598.

[26] Gupta R, Gopal R, Jat ML, Jat RK, Sidhu HS, Minhas PS, Malik RK. Wheat productivity in indo-gangetic plains of India during 2010: terminal heat effects and mitigation strategies. PACA Newslett. 2010; 14:1-11.

[27] Asseng S, Foster I, Turner NC. The impact of temperature variability on wheat yields. Glob. Change Biol. 2011; 17:997-1012.

[28] Lobell DB, Sibley A, Ortiz-Monasterio JI. Extreme heat effects on wheat senescence in India. Nat. Clim. Change. 2012; 2:186-189.

[29] Bell G, Collins S. Adaptation, extinction and global change. Evol. Appl. 2008; 1:3-16.

[30] Kelly AE, Goulden ML. Rapid shifts in plant distribution with recent climate change. Proc. Natl. Acad. Sci. USA 2008; 105:11823-11826.

[31] Shanthi-Prabha V, Sreekanth NP, Babu PK, Thomas AP. 2011. The trilemma of soil carbon degradation, climate change and food insecurity. Disaster, Risk and Vulnerability Conference 2011. School of Environmental Sciences, Mahatma Gandhi University, India - The Applied Geoinformatics for Society and Environment, Germany; 2011. Pp. 107-112.

[32] Gregory PJ, Johnson SN, Newton AC, Ingram JSI. Integrating pests and pathogens into the climate change/food security debate. J. Exp. Bot. 2009; 60:2827-2838.

[33] Patz JA, Kovats RS. Hot spots in climate change and human health: present and future risks. Lancet 2002; 368:859-869.

[34] McMichael, A, Woodruff RE, Hales S. Climate change and human health: present and future risks. Lancet 2006; 367:859-869.

[35] Ziska LH, Epstein PR, Schlesinger WH. Rising CO_2, climate change, and public health: exploring the links to plant biology. Environ. Health Perspect. 2009; 117: 155-158.

[36] Borlaug N. Contributions of conventional plant breeding to food production. Science 1983; 219:689-693.

[37] Trethowan RM, Reynolds MP, Ortiz-Monasterio I, Ortiz R. The genetic basis of the Green Revolution in wheat production. Plant Breed. Rev. 2007; 28:39-58.

[38] Burney JA, Davis SJ, Lobell DB. Greenhouse gas mitigation by agricultural intensification. Proc. Natl. Acad. Sci. USA 2010; 107:12052-12057.

[39] Edgerton MD. Increasing crop productivity to meet global needs for feed, food, and fuel. Plant Physiol. 2009; 149:7-13.

[40] Tilman D, Cassman KG, Matson PA, Naylor R, Polasky S. Agricultural sustainability and intensive production practices. Nature 2002; 418:671-677.

[41] IAASTD. Agriculture at the crossroads. International Assessement of Agricultural Knowledge, Science and Technology for Development (IAASTD). Island Press, Washington, DC; 2009.

[42] Motley TJ, Zerega N, Cross H. 2006. Darwin's harvest. New approach to the origins, evolution and conservation of crops. Columbia University Press, New York; 2006.

[43] Dias JS. Impact of improved vegetable cultivars in overcoming food insecurity. Euphytica 2010; 176:125-136.

[44] FAO. World summit on sustainable development. United Nations, August 29. United Nations, New York; 2002.

[45] Fowler C. Conserving diversity: the challenge of cooperation. Acta Hort. 2011; 916:19-24.

[46] Ortiz R. Not just seed repositories: a more proactive role for gene banks. In: Nordic Gene Bank 1979-1999. Nordic Gene Bank, Alnarp; 1999. Pp 45-49.

[47] SWPGRFA. Draft second report on the state of world plant genetic resources for food and agriculture (CGRFA-12/09/Inf.rRev.1). Twelth Regular Session, 19-23 Oct 2009, Rome, Italy; 2009.

[48] Reddy LJ, Kameswara-Rao N, Bramel PJ, Ortiz R. *Ex situ* genebank management at ICRISAT. In: Bhag M, Mathur PN, Ramantha-Rao V, Sajise PE (Eds.). Proceedings Fift Meeting of South Asia Network on Plant Genetic Resources, New Delhi, India, 9-11 October 2000. International Plant Genetic Resources Institute, South Asia Office, New Dehli; 2002. Pp 77-85.

[49] Fowler C, Hodgkin T. Plant genetic resources for food and agriculture: assessing global availability. Ann. Rev. Environ. Resour. 2004; 29:143-179.

[50] Leckie D, Astley D, Crute IR, Ellis PR, Pink DAC, Boukema I, Monteiro AA, Dias JS. The location and exploitation of genes for pest and disease resistance in European gene bank collections of horticultural brassicas. Acta Hort. 1996; 407:95-101.

[51] Santos MR, Dias JS. Evaluation of a core collection of *Brassica oleracea* accessions for resistance to white rust of crucifers (*Albugo candida*) at the cotyledon stage. Genetic Resources and Crop Evolution 2004; 51:713-722.

[52] Dias JS, Nogueira P, Corvo L. Evaluation of a core collection of *Brassica rapa* vegetables for resistance to *Xanthomonas campestris* pv. *campestris*. African Journal of Agricultural Research 2010; 5:2972-2980.

[53] Grenier C, Bramel PJ, Hamon J, Chantereau J, Deu M, Noirot M, Reddy VG, Kresovich S, Prasada-Rao KE, Mahalakshmi V, Crouch JH, Ortiz R. Core collections, DNA markers and bio-informatics: new tools for "mining" plant genetic resources held in

gene-banks - sorghum as an example. In: Oono K, Komatsuda T, Kadowaki K, Vaughan D (Eds.). Integration of biodiversity and Genome Technology for Crop Improvement. National Institute of Agro-Biological Resources, Tsukuba, Japan; 2000. Pp 139-140.

[54] Lenné JM, Ortiz R. Agrobiodiversity in pest management. In: Leslie JF (Ed.). Sorghum and Millet Diseases III. Iowa State University Press, Ames; 2002. Pp 309-320.

[55] Maxted N, Kell S, Ford-Lloyd B, Dulloo E, Toledo A. 2012. Toward the systematic conservation of global crop wild relative diversity. Crop Sci. 2012; 52: 774-785.

[56] Dwivedi SL, Upadhyaya HD, Stalker HT, Blair MW, Bertioli DJ, Nielen S, Ortiz R. Enhancing crop gene pools with beneficial traits using wild relatives. Plant Breed. Rev. 2008; 30:179-230.

[57] Jarvis A, Lane L, Hijmas R. The effect of climate changes on crop wild relatives. Agric. Ecosyst. Environ. 2008; 126:13-33.

[58] Ortiz R. Agrobiodiversity management for climate change. In: Lenné JM, Wood D (Eds.). Agrobiodiversity Management for Food Security, CAB International, Wallingford, Oxon, United Kingdom; 2011. Pp 189-211.

[59] Meilleur A, Hodgkin T. *In situ* conservation of crop wild relatives: status and trends. Biodivers. Conserv. 2004; 13:663-684.

[60] Maxted N, Dulloo E, Ford-Lloyd BV, Iriondo JM, Jarvis A. Gap analysis: a tool for complementary genetic conservation assessment. Divers. Distrib. 2008; 14: 1018-1030.

[61] Maxted N, Kell SP. 2009. Establishment of a global network for the *in situ* conservation of crop wild relatives: status and needs. FAO Commission on Genetic Resources for Food and Agriculture, FAO, Rome: 2009. Pp 266.

[62] Rana JC, Sharma SK. Plant genetic resource management under emerging climate change. Indian J. Genet. 2009; 69:1-17.

[63] Wardle DA. Communities and ecosystems: linking the aboveground and belowground components. Princeton University Press, Princeton NJ, USA; 2002.

[64] Paoletti MG, Foissner W, Coleman D. 1994. Soil biota, nutrient cycling, and farming systems. Lewis Publishers, Boca Raton, Florida; 1994.

[65] Altieri MA. The ecological role of biodiversity in agroecosystems. Agric. Ecosyst. Environ. 1999; 74:19-31.

[66] Roem WJ, Berendse F. Soil acidity and nutrient supply ratio as possible factors determining changes in plant species diversity in grassland and heathland communities. Biological Conservation 2000; 92(2):151-161.

[67] Wall DH, Bardgett RD, Kelly EF. Biodiversity in the dark. Nat. Geosci. 2010; 3:297-298.

[68] Sylvain ZA, Wall DH. Linking soil biodiversity and vegetation: implications for a changing planet. Am. J. Bot. 2011; 98:517-527.

[69] Prichard SG. Soil organisms and global climate change. Plant Pathol. 2011; 60:82-99.

[70] Castro HF, Classen AT, Austin EE, Norby RJ, Schadt CW. Soil microbial community responses to multiple experimental climate change drivers. Appl. Environ. Microbiol. 2010; 76(4):999-1007.

[71] Dominique G, Schmid B, Brandl H. Influence of plant diversity and elevated atmospheric carbon dioxide levels on belowground bacterial diversity. BMC Microbiol. 2006; 6:68.

[72] Bardgett RD. Plant-soil interactions in a changing world. F1000 Rep. Biol. 2011; 3:16.

[73] Zhili H, Xu M, Deng Y, Kang S, Kellog L, Wu L, van Nostrand JD, Hobbie SE, Reich PB, Zhou J. Metagenomic analysis reveals a marked divergence in the structure of belowground microbial communities at elevated CO_2. Ecol. Lett. 2010; 13: 564-575.

[74] Deng Y, He Z, Xu M, Qin Y, van Nostrand JD, Wu L, Roe BA, Wiley G, Hobbie SE, Reich PB, Zhou J. Elevated carbon dioxide alters the structure of soil microbial communities. Appl. Environ. Microbiol. 2012; 78:2991-2995.

[75] Frison E, Cherfas J, Hodgkin T. Agricultural biodiversity is essential for a sustainable improvement in food and nutrition security. Sustainability 2011; 3:238-253.

[76] Tilman D, Reich PB, Knops J, Wedin D, Mielke T, Lehman C. Diversity and productivity in a long-term grassland experiment. Science 2001; 294:843-845.

[77] Bullock JM, Pywell RF, Walker KJ. Long-term enhancement of agricultural production by restoration of biodiversity. J. Appl. Ecol. 2007; 44:6-12.

[78] Proulx R, Wirth C, Voigt W, Weigelt A, Roscher C, Attinger S, Baade J, Barnard RL, Buchmann N, Buscot F, Eisenhauer N, Fischer M, Gleixner G, Halle S, Hildebrandt A, Kowalski E, Kuu A, Lange M, Milcu A, Niklaus PA, Oelmann Y, Rosenkranz S, Sabais A, Scherber C, Scherer-Lorenzen M, Scheu S, Schulze E-D, Schumacher J, Schwichtenberg G, Soussana J-F, Temperton VM, Weisser WW, Wilcke W, Schmid B. Diversity promotes temporal stability across levels of ecosystem organization in experimental grasslands. Public Library of Science (PLoS) One 2010; 5:e13382.

[79] Finckh MR, Gacek ES, Goyeau H, Lannou C, Merz U, Mundt CC, Munk L, Nadziak J, Newton AC, De Vallavielle-Pope C, Wolfe MS. Cereal variety and species mixtures in practice, with emphasis on disease resistance. Agronomie 2000; 20:813-837.

[80] Eyzaguirre PB, Linares OF. Home gardens and agro-biodiversity. Smithsonian, Washington, DC, USA; 2004.

[81] Galluzzi G, Eyzaguirre P, Negri V. Home gardens: neglected hotspots of agro-biodiversity and cultural diversity. Biodivers. Conserv. 2010; 19:3635-3654.

[82] Harlan JR. Crops and man. American Society of Agronomy and Crop Science Society of America. Madison, WI, USA; 1992.

[83] Pratt RC. A historical examination of the development and adoption of hybrid corn: a case study in Ohio. Maydica 2004; 49:155-172.

[84] Dias JS. Biodiversity and vegetable breeding in the light of developments in intellectual property rights. In: Grillo O,Verona G (Eds.). 17. Ecosystems Biodiversity. IN-TECH Publ., Rijeka, Croatia; 2011. Pp 389-428.

[85] Dias JS. Impact of the vegetable breeding industry and intellectual property rights in biodiversity and food security. In: Jones AM, Hernandez FE (Eds.). Food security: quality, management, issues and economic implications. Nova Science Publishers Inc., Hauppauge, New York, USA; 2012. Pp 57-86.

[86] Dias JS. Impact of vegetable breeding industry and intellectual property rights in food security. In: Nath P (Ed.). The Basics of Human Civilization-Food, Agriculture and Humanity, Volume-I-Present Scenario. Prem Nath Agricultural Science Foundation (PNASF), Bangalore & New India Publishing Agency (NIPA), New Delhi, India; 2013. Pp173-198.

[87] Dias JS, Ryder E. Impact of plant breeding on the world vegetable industry. Acta Hortic. 2012; 935:13-22.

[88] Dias JS. 23. Impact of improved vegetable cultivars in overcoming food insecurity. In: Nath P, Gaddagimath PB (Eds.). Horticulture and Livelihood Security. Scientific Publishers, New Dehli, India; 2010. Pp 303-339.

[89] Dias JS, Ortiz R. Transgenic vegetable crops: progress, potentials and prospects. Plant Breeding Reviews 2012; 35:151-246.

[90] Dias JS. The use of molecular markers in selection of vegetables. SECH, Actas Horticultura 1989; 3:175-181.

[91] Dias JS. The use of computers in plant breeding. SECH, Actas Horticultura, 1991; 8:367-371.

[92] Dias JS, Ortiz R. Transgenic vegetable breeding for nutritional quality and health benefits. Food Nutr. Sci. 2012; 3(9):1209-1219.

[93] Dias JS, Ortiz R. Transgenic vegetables for Southeast Asia. In: Holmer R, Linwattana G, Nath P, Keatinge JDH (Eds). Proc. Regional Symposium on High Value Vegetables in Southeast Asia: Production, Supply and Demand (SEAVEG 2012). January 2012, Chiang Mai, Thailand. AVRDC - The World Vegetable Center, Publication No. 12-758. AVRDC, Shanhua, Tainan,Taiwan; 2013. Pp 361-369.

[94] Dias JS, Ortiz R. Transgenic vegetables for 21st century horticulture. Acta Hortic. 2013; 974:15-30.

[95] Tilman D. Global environmental impacts of agricultural expansion: the need for sustainable and efficient practices. Proc. Natl. Acad. Sci. USA 1999; 96:5995-6000.

[96] Robertson GP, Swinton SM. Reconciling agricultural productivity and environmental integrity: a grand challenge for agriculture. Front. Ecol. Environ. 2005; 3: 38-46.

[97] Hirel B, Le Gouis J, Ney B, Gallais A. The challenge of improving nitrogen use efficiency in crop plants: towards a more central role for genetic variability and quantitative genetics within integrated approaches. J. Exp. Bot. 2007; 58: 2369-2387.

[98] Foulkes MJ, Hawkesford MJ, Barraclough PB, Holdsworth MJ, Kerr S, Kightley S, Shewry PR. Identifying traits to improve the nitrogen economy of wheat: recent advances and future prospects. Field Crop Res 2009; 114:329-342.

[99] Korkmaz K, Ibrikci H, Karnez E, Buyuk G, Ryan J, Ulger AC, Oguz H. Phosphorus use efficiency of wheat genotypes grown in calcareous soils. J. Plant Nutr. 2009; 32:2094-2106.

[100] Farooq M, Kobayashi N, Wahid A, Ito O, Basra SMA. Strategies for producing more rice with less water. Advances in Agronomy 2009; 101:351-388.

[101] Shi W, Moon CD, Leahy SC, Kang D, Froula J, Kittelmann S, Fan C, Deutsch S, Gagic D, Seedorf H, Kelly WJ, Atua R, Sang C, Soni P, Li D, Pinares-Patiño CS, McEwan JC, Janssen PH, Chen F, Visel A, Wang Z, Attwood GT, Rubin EM. Methane yield phenotypes linked to differential gene expression in the sheep rumen microbiome. Genome Res. 2014; 24:1517-1525.

[102] Shrawat AK, Good AG. Genetic engineering approaches to improving nitrogen use efficiency. Plant Research News. ISB Report May 2008. Information Systems for Biotechnology (ISB) News Report, Blackburg, VA, USA; 2008. http://www.isb.vt.edu/news/2008/news08.may.htm#may0801

[103] Daemrich A, Reinhardt F, Shelman M. Arcadia biosciences: seeds of change. Harvard Business School, Boston, Massachusetts, USA; 2008.

[104] Subbarao GV, Ban T, Kishii M, Ito O, Samejima HY, Wang SJ, Pearse S, Gopalakrishnan K, Nakahara AKM, Zakir Hossain H, Tsujimoto WL, Berry W. Can biological nitrification inhibition (BNI) genes from perennial *Leymus racemosus* (*Triticeae*) combat nitrification in wheat farming? Plant Soil 2007; 299:55-64.

[105] IPCC (Intergovernmental Panel on Climate Change). The physical science basis. In: Solomon S, Qin D, Manning M, Chen Z, Marquis M, Averyt KB, Tignor M, Miller HL (Eds). Contribution of Working Group I to the Fourth Assessment Report of the Intergovernmental Panel on Climate Change. Cambridge University Press, Cambridge, MA, USA; 2009.

[106] Ortiz R, Sayre KD, Govaerts B, Gupta R, Subbarao GV, Ban T, Hodson D, Dixon JM, Ortiz-Monasterio JI, Reinolds M. Climate change: can wheat beat the heat? Agr. Ecosyst. Environ. 2008; 126:46-58.

[107] Araus J, Slafer G, Royo C, Serret MD. Breeding for yield potential and stress adaptation in cereals. Crit. Rev. Plant Sci. 2008; 27:377-412.

[108] Cattivelli L, Rizza F, Badeck FW, Mazzucoteli E, Mastrangelo AM, Francia E, Marè C, Tondelli A, Stanca AM. Drought tolerance improvement in crop plants: an integrated view from breeding to genomics. Field Crop Res 2008; 105(1-2):1-14.

[109] Ceccarelli S, Grando S. Decentralized-participatory plant breeding: an example of demand driven research. Euphytica 2007; 155:349-360.

[110] Burke MB, Lobell DB, Guarino L. Shifts in African crop climates by 2050, and the implications for crop improvement and genetic resources conservation. Glob. Environ. Chang. 2009; 19:317-325.

[111] Jarvis DI, Brown AHD, Cuong PH, Collado-Panduro L, Latoumerie-Moreno L, Gyawali S, Tanto T, Sawadogo M, Mar I, Sadiki M, Hue NT, Arias-Reyes L, Balma D, Bajracharya J, Castillo F, Rijal D, Belqadi L, Rana R, Saidi S, Quedraogo J, Zangre R, Rhrib K, Chavez JL, Schoen D, Shapit B, Santis PD, Fadda C, Hodgkin T. A global perspective of the richness and evenness of traditional crop-variety diversity maintained by farming communities. Proc. Natl. Acad Sci. USA 2008; 105(14):5326-5331.

[112] Witcombe JR, Hollington PA, Howarth CJ, Reader S, Steele KA. Breeding for abiotic stresses for sustainable agriculture. Philos. Trans. Roy. Soc. Lon. B. Biol. Sci. 2008; 363:703-716.

[113] Bhatnagar-Mathur P, Vadez V, Sharma KK. Transgenic approaches for abiotic stress tolerance in plants: retrospect and prospects. Plant Cell Reports, 2007; 27(3):411-424.

[114] Ainsworth E, Rogers A, Leakey ADB. Targets for crop biotechnology in a future high-CO2 and high-O3 world. Plant Physiology 2008; 147:13-19.

[115] Ortiz R. Crop genetic engineering under global climate change. Annals Arid Zone 2008; 47:343-354.

[116] Jewell MC, Campbell BC, Godwin ID. Transgenic plants for abiotic stress resistance. In: Kole C, Michler CH, Abbott AG, Hall TC (Eds.). Transgenic Crop Plants. Springer-Verlag, Berlin-Heidelberg, Germany; 2010.

[117] Dwivedi SL, Upadhyaya H, Subudhi P, Gehring C, Bajic V, Ortiz R. Enhancing abiotic stress tolerance in cereals through breeding and transgenic interventions. Plant Breeding Reviews 2010; 33:31-114.

[118] Dwivedi SL, Sahrawat K, Upadhyaya H, Ortiz R. Food, nutrition and agrobiodiversity under global climate change. Advances in Agronomy 2013; 120:1-118.

[119] Ruane J, Sonnino A, Steduto P, Deane C. Coping with water scarcity: What role for biotechnologies? Land and Water Discussion Paper 7. Food and Agriculture Organization of the United Nations, Rome, Italy; 2008.

[120] Sakuma Y, Maruyama K, Qin F, Osakabe Y, Shinozaki K, Yamaguchi-Shinozaki K. Dual function of an *Arabidopsis* transcription factor DREB2A in water-stress-respon-

sive and heat-stress-responsive gene expression. Proc. Natl. Acad. Sci. USA 2006; 103:18822-18827.

[121] Ortiz R, Iwanaga M, Reynolds MP, Wu X, Crouch JH. Overview on crop genetic engineering for drought-prone environments. J. Semi-Arid Trop. Agric. Res. 2007; 4(1): 1-30.

[122] Pellegrineschi A, Reynolds M, Pacheco M, Brito RM, Almeraya R, Yamaguchi-Shinozaki K, Hoisington D. Stress-induced expression in wheat of the *Arabidopsis thaliana* DREB1A gene delays water stress symptoms under greenhouse conditions. Genome 2004; 47: 493-500.

[123] Saint Pierre CS, Crossa JL, Bonnett D, Yamaguchi-Shinozaki K, Reynolds MP. Phenotyping transgenic wheat for drought resistance. J. Exp. Bot. 2012; 63:1799-1808.

[124] Mumms R. Genes and salt tolerance: bringing them together. New Phytologist. 2005; 167:645-663.

[125] Chinnusamy V, Jagendorf A, Zhu J-K. Understanding and improving salt tolerance in plants. Crop Science 2005; 45:437-448.

[126] Wu HJ, Zhang Z, Wang JY, Oh DH, Dassanayake M, Liu B, Huang Q, Sun HX, Xia R, Wu Y, Wang YN, Yang Z, Liu Y, Zhang W, Zhang H, Chu J, Yan C, Fang S, Zhang J, Wang Y, Zhang F, Wang G, Yeol Lee S, Cheeseman JM, Yang B, Li B, Min J, Yang L, Wang J, Chu C, Chen SY, Bohnert HJ, Zhu JK, Xiu-Jie Wang XJ, Xiea Q. Insights into salt tolerance from the genome of *Thellungiella salsuginea*. PNAS, 2012; 109:12219-12224.

[127] Plett D, Safwat G, Gilliham M, Skrumsager-Møller I, Roy S, Shirley N, Jacobs A, Johnson A, Tester M. Improved salinity tolerance of rice through cell type-specific expression of AtHKT1;1. PLoS ONE 2010; 5(9):e12571.

[128] Moghaieb RE, Nakamura A, Saneoka H, Fujita K. Evaluation of salt tolerance in ectoine-transgenic tomato plants (*Lycopersicon esculentum*) in terms of photosynthesis, osmotic adjustment, and carbon partitioning. GM Crops 2011; 2:58-65.

[129] Lybbert T, Sumner D. Agricultural technologies for climate change mitigation and adaptation in developing countries: policy options for innovation and technology diffusion. ICTSD-IPC Platform on Climate Change. Agriculture and Trade Issues Brief 6. International Centre for Trade and Sustainable Development, Geneva, Switzerland - International Food & Agricultural Trade Policy Council, Washington DC; 2011.

[130] Varshney RK, Bansal KC, Aggarwal PK, Datta SK, Craufurd PQ. Agricultural biotechnology for crop improvement in a variable climate: hope or hype? Trends Plant Sci. 2011; 16:363-371.

[131] Yamori W, Hikosaka K, Way DA. Temperature response of photosynthesis in C3, C4, and CAM plants. Photosynthesis Res. 2013; 119(1-2):101-117.

[132] Ainsworth EA, Ort DR. 2010. How do we improve crop production in a warming world? Plant Physiol. 2010; 154:526-530.

[133] Wahid A, Gelani S, Ashraf M, Foolad MR. Heat tolerance in plants: An overview. Experim. Bot. 2007; 61:199-223.

[134] Hasanuzzaman M, Nahar K, Alam MdM, Roychowdhury R, Fujita M. Physiological, biochemical, and molecular mechanisms of heat stress tolerance in plants. Intern. J. Molec. Sci. 2013; 14:9643-9684.

[135] Gao H, Brandizzi F, Benning C, Larkin RM. A membrane-tethered transcription factor defines a branch of the heat stress response in Arabidopsis thaliana. Proc. Natl. Acad. Sci. USA 2008; 105:16399-16404.

[136] Gusta L. Abiotic stresses and agricultural sustainability. J. Crop Improv. 2012; 26:415-427.

[137] Katiyar-Agarwal S, Agarwal M, Grover A. Heat-tolerant basmati rice engineered by over-expression of hsp101. Plant Molec. Biol. 2003; 51:677-686.

[138] Pimentel D. Techniques for reducing pesticide use. Economic and environmental benefits. Wiley, New York; 1997.

[139] Oerke EC, Dehne HW, Schonbeck F, Weber A. Crop production and crop protection: Estimated losses in major food and cash crops. Elsevier, Amsterdam; 1994.

[140] Tripathi L, Mwaka H, Tripathi JN, Tushemereirwe W. Expression of sweet pepper Hrap gene in banana enhances resistance to *Xanthomonas campestris* pv *musacearum*. Molecular Plant Pathol. 2010; 11:721-731.

[141] Ortiz R, Jarvis A, Aggarwal PK, Campbell BM. Plant genetic engineering, climate change and food security. CCAFS Working Paper No. 72. CGIAR Research Program on Climate Change, Agriculture and Food Security (CCAFS). Copenhagen, Denmark; 2014.

[142] Haverkort AJ, Boonekamp PM, Hutten R, Jacobsen E, Lotz LAP, Kessel GJT, Visser RFG, van der Vossen EAG. Societal costs of late blight in potato and prospects of durable resistance through cisgenic modification. Potato Res. 2008; 51:47-57.

[143] Jacobs DF. Toward development of silvical strategies for forest restoration of American chestnut (*Castanea dentata*) using blight resistant hybrids. Biol. Conserv. 2007; 137:497-506.

[144] Santini A, La Porta N, Ghelardini L, Mittempergher L. Breeding against Dutch elm disease adapted to the Mediterranean climate. Euphytica 2007; 163:45-56.

[145] Bouton J. The economic benefits of forage improvement in the United States. Euphytica 2007; 154:263-270.

[146] Pimentel D, Allen J, Beers A, Guinand L, Linder R, McLaughlin P, Meer B, Musonda D, Perdue D, Poisson S, Siebert S, Stoner K, Salazar R, Hawkinset A. World agricul-

ture and soil erosion. Erosion threatens world food production. BioScience 1987; 37(4):277-283.

[147] Glover JD, Cox CM, Reganold JP. Future farming: a return to roots? Sci. Amer. 2007; 297:82-89.

[148] Jackson W, Cox S, DeHaan L, Glover J, Van Tassel D, Cox C. The necessity and possibility of an agriculture where nature is the measure. In: Bohlen PJ, House G (Eds.). Sustainable agroecosystem management. CRC Press, Boca Raton, FL, USA; 2009.

[149] Blanco H, Lal R. (Eds.). Principles of soil conservation and management. Springer, New York; 2008.

[150] Olson KR, Ebelhar SA, Lang JM. Cover crops effects on crop yields and soil organic content. Soil Sci 2010; 175(2):89-98.

[151] Kaumbutho P, Kienzle J. Conservation agriculture as practiced in Kenya: two case studies. Food and Agriculture Organization of the United Nations, Rome; 2008.

[152] Kendle AD, Rose JE. The aliens have landed! What are the justifications for "native only" policies in landscape plantings? Landscape Urban Plan 2000; 47:19-31.

[153] Zhao FJ, McGrath SP. Biofortification and phytoremediation. Curr. Opin. Plant Biol. 2009; 12:373-380.

[154] Yin X, Yuan L, Liu Y, Lin Z. 2012. Phytoremediation and biofortification:two sides of one coin. In: Yin X, Yuan L. (Eds.). Phytoremediation and Biofortification. Springer-Briefs in Green Chemistry for Sustainable. Springer, New York; 2012. Pp 1-6.

[155] McGrath SP, Lombi E, Gray CW, Caille N, Dunham SJ, Zhao FJ. Field evaluation of Cd and Zn phytoextraction potential by the hyperaccumulators *Thlaspi caerulescens* and *Arabidopsis halleri*. Environ. Pollut. 2006; 141:115-125.

[156] Maxted AP, Black CR, West HM, Crout NMJ, McGrath SP,Young SD. Phytoextraction of cadmium and zinc from arable soils amended with sewage sludge using *Thlaspi caerulescens*: development of a predictive model. Environ. Pollut. 2007; 150:363-372.

[157] Kertulis-Tartar GM, Ma LQ, Tu C, Chirenje T. Phytoremediation of an arsenic-contaminated site using *Pteris vittata* L. A two-year study. Int. J. Phytoremed. 2006; 8:311-322.

[158] Salido AL, Hasty KL, Lim JM, Butcher DJ. Phytoremediation of arsenic and lead in contaminated soil using Chinese brake ferns (*Pteris vittata*) and Indian mustard (*Brassica juncea*). Int. J. Phytoremed. 2003; 5:89-103.

[159] Cook RJ. Toward cropping systems that enhance productivity and sustainability. Proc. Natl. Acad. Sci. USA 2006; 103:18389-18394.

[160] Ortiz R, Mowbray D, Dowswell C, Rajaram S. Norman E. Borlaug: The humanitarian plant scientist who changed the world. Plant Breed. Rev. (2007) 28:1-37.

[161] Ortiz R, Braun HJ, Crossa J, Crouch JH, Davenport G, Dixon J, Dreisigacker S, Duveiller E, He Z, Huerta J, Joshi AK, Kishii M, Kosina P, Manes Y, Mezzalama M, Morgounov A, Murakami J, Nicol J, Ortiz-Ferrara G, Ortiz-Monasterio JI, Payne TS, Peña RJ, Reynolds MP, Sayre KD, Sharma RC, Singh RP, Wang J, Warburton M, Wu H, Iwanaga M.Wheat genetic resources enhance- ment by the International Maize and Wheat Improvement Center (CIMMYT). Genet. Resour. Crop Evol. 2008; 55:1095-1140.

[162] Reynolds MP, Borlaug NE. (2006). International collaborative wheat improvement: Impacts and future prospects. J. Agric. Sci. 2006; 144:3-17.

[163] Lantican MA, Dubin MJ, Morris ML. Impacts of international wheat breeding research in the developing world, 1988-2002. Centro Internacional de Mejoramiento de Maíz y Trigo, México D.F.; 2005.

[164] Alston JM, Marra MC, Pardey PG, Wyatt TJ. Research returns redux: A meta-analysis of the returns to agricultural R&D. Austr. J. Agric. Resour. Econ. 2000; 44: 185-215.

[165] Evenson RE, Gollin D. Genetic resources, international organizations, and improvement in rice varieties. Econ. Dev. Cultural Change 1997; 45:471-500.

[166] Jackson MT, Huggan RD. Sharing the diversity of rice to feed the world. Diversity 1993; 9:22-25.

[167] Salhuana W, Pollak L. Latin American Maize Project (LAMP) and Germplasm Enhancement of Maize (GEM) Project: generating useful breeding germplasm. Maydica 2006; 51: 339-355.

[168] Taba S, Díaz J, Franco J, Crossa J, Eberhart SA. A core subset of LAMP from the Latin American Maize Project. CD-Rom. Centro Internacional de Mejoramiento de Maíz y Trigo, México D.F.; 1999.

[169] Balint-Kurti P, Blanco M, Milard M, Duvick S, Holland J, Clements M, Holley R, Carson ML, Goodman M. Registration of 20 GEM maize breeding germplasm lines adapted to the southern U.S. Crop Sci. 2006; 46:996-998.

[170] Goodman MM. Broadening the U.S. maize germplasm base. Maydica 2005; 50:203-214.

[171] Ortiz R, Taba S, Chávez-Tovar VH, Mezzalama M, Xu Y, Yan J, Crouch JH. Conserving and enhancing maize genetic resources as global public goods - A perspective from CIMMYT. Crop Sci. 2010; 50:13-28.

[172] Delmer DP. Agriculture in the developing world: connecting innovations in plant research to downstream applications. Proc. Natl. Acad. Sci. USA 2005; 102:15739-15746.

[173] Conner AJ, Mercer CF. Breeding for success: diversity in action. Euphytica 2007; 154:261-262.

[174] Huang J, Pray C, Rozelle S. Enhancing the crops to feed the poor. Nature 2002; 418:678-684.

[175] Welch RM, Graham RD. Breeding for micronutrients in staple food crops from a human nutrition perspective. J. Exp. Bot. 2004; 55:353-364.

[176] Lichtfouse E, Navarrete M, Debaeke P, Souchere V, Alberola C, Menassieu J. Agronomy for sustainable agriculture. A review. Agron. Sustain. Dev. 2009; 29:1-6.

[177] Lal R. Soils and sustainable agriculture. A review. Agron. Sustain. Dev. 2008; 28:57-64.

[178] Pingali P, Raney T. (2005) From the green revolution to the gene revolution: how will the poor fare? ESA working paper no. 05- 09, November 2005. Agricultural and Development Economics Division (ESA). FAO, Rome, Italy; 2005. 15 pp.

[179] Jain HK. Green revolution: history, impact and future. Studium Press, Housten; 2010.

8

Characterizing *Tabebuia rosea* (Bertol.) DC. Using Microsatellites in Provenance and Progeny Trials in Colombia

Ana María López, Marta Leonor Marulanda and
Carlos Mario Ospina

1. Introduction

The area under forests is estimated to be approximately 3,870 million hectares worldwide, 95% of which corresponds to natural forests or native woodlands and the remaining 5% to planted forests [1]. Forests are dual purpose—commercial exploitation and environmental improvement [2]. Because of the pressure exerted by environmentalists, current global efforts strive to reduce timber extraction in natural forests. Forecasts are that the future increase in demand for timber will be covered by trees specifically planted for this purpose. Because of their capacity to fix carbon, forests will play an increasingly important role in view of current climate change attributed to greenhouse gas emissions [3].

Genetic improvement programs for forest species face diverse problems such as the long regeneration periods and the high costs involved in maintaining a forest population over a long period of time at different locations. Furthermore, compared with short-cycle crops, the returns to investment in forest plantations are definitively more delayed [4]. Molecular markers have a great impact on genetic improvement because they minimize the intervals of regeneration, increase the genetic gain per generation, and accumulate genetic information key for 'non-domesticated' species. Molecular markers are increasingly included in forest genetic breeding programs, where they are used to estimate polymorphism, establish parameters of relationship and mating systems, characterize genotypes, and assist in selection processes [5]. Microsatellites are the markers that have been most used in recent years and can be used to design seed orchards to estimate the contamination of pollen from external sources and to study mating models and variation of male fertility.

The development of microsatellite markers for forest species has been limited. The ratio of use of this type of marker in forest species, as compared its use in other cultivated species, is 6:1. Exceptions to the above are forest species used in the timber industry, such as *Pinus, Quercus,* and *Eucalyptus* [6, 7].

In Latin America, Brazil plays a pioneering role in the development of microsatellite markers for forest species of economic and ecological importance, for example *Caryocar brasiliense* [8], *Ceiba pentandra* [9], *Copaifera langsdorffii* [10], *Eugenia uniflora* [11], and several *Cariniana* species [12], among others.

In Colombia, coffee has been traditionally grown in association with forest species such as *Tabebuia rosea* and *Cordia alliodora*, which serve as shade, live fences, or perimeters [13]. Their characteristics, such as good wood quality, make them very appropriate to be used in agroforestry systems. *Tabebuia rosea,* a species with high-value, good-quality wood, is very important in Central America. Its lilac-pink flowers make it one of the most eye-catching trees of Central and South America, where it is mainly used as a shade and ornamental plant because of the beauty of its pink flower panicles [14, 15] (Figure 1). Work carried out with microsatellites for the *Tabebuia* genus includes studies carried out in *Tabebuia aurea* by Braga et al. [16], who developed 21 polymorphic microsatellites using a genomic library. This current study characterized, using microsatellite molecular markers, the best-performing *T. rosea* materials of provenance and progeny trials established by Colombia's National Center for Coffee Research (CENICAFE) in the country's coffee-growing region as well as the best materials established by the Santa Rosalía de Palermo Reforestation Company (REFOPAL) and the National Corporation for Forest Research and Promotion (CONIF) in the country's Caribbean region.

Figure 1. *Tabebuia rosea* tree in Colombia's coffee region

2. Materials and methods

2.1. Plant material

Samples were taken from provenance and progeny trials carried out by CENICAFE, a REFOPAL-CONIF clonal orchard, and a CONIF progeny orchard. The provenance and progeny trial, established by CENICAFE in 1997, is located in Colombia's coffee-growing region. Seed from plus trees collected in Colombia as well as in El Salvador, Guatemala, and Nicaragua were introduced for this trial (Table 1).

The REFOPAL-CONIF clonal seed orchard for conservation purposes, established in 1999, is located in Colombia's Caribbean region (code SA). Several progeny of this trial are also located in the Caribbean region (code HI). The REFOPAL-CONIF trial involves 24 sites of origin, 21 of which correspond to clones originating from plus trees and 3 from commercial seed (absolute controls) (codes HIT1, HIT2, and HIT3) (Table 1). All samples were identified according to their origin: Population 1, REFOPAL and CONIF clonal orchards; Population 2, progenies of REFOPAL and CONIF clonal orchards; and Population 3, CENICAFE provenance and progeny trials.

Progeny/clone/plus tree (no.)	Code	Country of origin	Type of trial	Trial or collection site (plus trees)
1	SA2	Colombia (Atlántico)	Clonal orchard	San Antero (Córdoba)
2	SA16	Colombia (Córdoba)	Clonal orchard	San Antero (Córdoba)
3	SA32	Colombia (Córdoba)	Clonal orchard	San Antero (Córdoba)
4	SA3	Colombia (Atlántico)	Clonal orchard	San Antero (Córdoba)
5	SA17	Colombia (Córdoba)	Clonal orchard	San Antero (Córdoba)
6	SA33	Colombia (Córdoba)	Clonal orchard	San Antero (Córdoba)
7	SA4	Colombia (Bolívar)	Clonal orchard	San Antero (Córdoba)
8	SA18	Colombia (Córdoba)	Clonal orchard	San Antero (Córdoba)
9	SA34	Colombia (Córdoba)	Clonal orchard	San Antero (Córdoba)
10	SA6	Colombia (Atlántico)	Clonal orchard	San Antero (Córdoba)
11	SA19	Colombia (Córdoba)	Clonal orchard	San Antero (Córdoba)
12	SA35	Colombia (Córdoba)	Clonal orchard	San Antero (Córdoba)
13	SA7	Colombia (Atlántico)	Clonal orchard	San Antero (Córdoba)
14	SA20	Colombia (Córdoba)	Clonal orchard	San Antero (Córdoba)
15	SA36	Colombia (Córdoba)	Clonal orchard	San Antero (Córdoba)
16	SA8	Colombia (Bolívar)	Clonal orchard	San Antero (Córdoba)
17	SA22	Colombia (Sucre)	Clonal orchard	San Antero (Córdoba)

Progeny/clone/plus tree (no.)	Code	Country of origin	Type of trial	Trial or collection site (plus trees)
18	SA38	Colombia (Córdoba)	Clonal orchard	San Antero (Córdoba)
19	SA10	Colombia (Magdalena)	Clonal orchard	San Antero (Córdoba)
20	SA24	Colombia (Sucre)	Clonal orchard	San Antero (Córdoba)
21	SA41	Colombia (Córdoba)	Clonal orchard	San Antero (Córdoba)
22	SA11	Colombia (Córdoba)	Clonal orchard	San Antero (Córdoba)
23	SA25	Colombia (Sucre)	Clonal orchard	San Antero (Córdoba)
24	SA42	Colombia (Córdoba)	Clonal orchard	San Antero (Córdoba)
25	SA12	Colombia (Córdoba)	Clonal orchard	San Antero (Córdoba)
26	SA27	Colombia (Sucre)	Clonal orchard	San Antero (Córdoba)
27	SA43	Colombia (Córdoba)	Clonal orchard	San Antero (Córdoba)
28	SA13	Colombia (Córdoba)	Clonal orchard	San Antero (Córdoba)
29	SA28	Colombia (Sucre)	Clonal orchard	San Antero (Córdoba)
30	SA44	Colombia (Córdoba)	Clonal orchard	San Antero (Córdoba)
31	SA14	Colombia (Córdoba)	Clonal orchard	San Antero (Córdoba)
32	SA29	Colombia (Sucre)	Clonal orchard	San Antero (Córdoba)
33	SA46	Colombia (Magdalena)	Clonal orchard	San Antero (Córdoba)
34	SA15	Colombia (Córdoba)	Clonal orchard	San Antero (Córdoba)
35	SA31	Colombia (Córdoba)	Clonal orchard	San Antero (Córdoba)
36	HI31	Colombia (Córdoba)	Progeny of clonal orchards	La Independencia Hacienda
37	HI9	Colombia (Magdalena)	Progeny of clonal orchards	La Independencia Hacienda
38	HI33	Colombia (Córdoba)	Progeny of clonal orchards	La Independencia Hacienda
39	HI35	Colombia (Córdoba)	Progeny of clonal orchards	La Independencia Hacienda
40	H1T2	Check	Commercial material 2	La Independencia Hacienda
41	HIT3	Check	Commercial material 3	La Independencia Hacienda
42	HI19	Colombia (Córdoba)	Progeny of clonal orchards	La Independencia Hacienda
43	HI47	Colombia (Sucre)	Progeny of clonal orchards	La Independencia Hacienda
44	H1T1	Check	Commercial material 1	La Independencia Hacienda

Progeny/clone/plus tree (no.)	Code	Country of origin	Type of trial	Trial or collection site (plus trees)
45	HI23	Colombia (Sucre)	Progeny of clonal orchards	La Independencia Hacienda
46	HI44	Colombia (Córdoba)	Progeny of clonal orchards	La Independencia Hacienda
47	HI5	Colombia (Bolívar)	Progeny of clonal orchards	La Independencia Hacienda
48	061/96	El Salvador	CENICAFE provenance and progeny trial	Floridablanca (Santander). Pueblo Bello (Cesar)
49	506/92	Guatemala	CENICAFE provenance and progeny trial	Floridablanca (Santander). Pueblo Bello (Cesar)
50	SO2386	Nicaragua	CENICAFE provenance and progeny trial	Floridablanca (Santander). Pueblo Bello (Cesar)
51	AGO95	Guatemala	CENICAFE provenance and progeny trial	Floridablanca (Santander). Pueblo Bello (Cesar)
52	CU-II-1*	Colombia (Cundinamarca)	CENICAFE provenance and progeny trial	Floridablanca (Santander). Pueblo Bello (Cesar)
53	MAR 96-09	Guatemala	CENICAFE provenance and progeny trial	Floridablanca (Santander). Pueblo Bello (Cesar)
54	AG 95-30	Guatemala	CENICAFE provenance and progeny trial	Floridablanca (Santander). Pueblo Bello (Cesar)
55	SO2386	Nicaragua	CENICAFE provenance and progeny trial	Floridablanca (Santander). Pueblo Bello (Cesar)
56	M-I-1*	Colombia	CENICAFE provenance and progeny trial	Floridablanca (Santander). Pueblo Bello (Cesar)
57	CU-I-1-10	Colombia	CENICAFE plus trees	Cundinamarca
58	CU-I-1-11	Colombia	CENICAFE plus trees	Cundinamarca
59	CU-I-1-12	Colombia	CENICAFE plus trees	Cundinamarca
60	CU-II-1-10	Colombia	CENICAFE plus trees	Cundinamarca
61	CU-II-1-11	Colombia	CENICAFE plus trees	Cundinamarca
62	ABR 95-24	Guatemala	CENICAFE plus trees	Floridablanca (Santander) Pueblo Bello (Cesar)
63	CU-II-*	Colombia	CENICAFE provenance and progeny trial	Floridablanca (Santander) Pueblo Bello (Cesar)
64	ABR 95-30	Guatemala	CENICAFE provenance and progeny trial	Floridablanca (Santander) Pueblo Bello (Cesar)

Progeny/clone/plus tree (no.)	Code	Country of origin	Type of trial	Trial or collection site (plus trees)
65	MAR96/09	Guatemala	CENICAFE provenance and progeny trial	Líbano (Tolima)
66	AGO25-30	Guatemala	CENICAFE provenance and progeny trial	Fredonia (Antioquia)
67	ABR 95-24	Guatemala	CENICAFE provenance and progeny trial	Fredonia (Antioquia)
68	AGO95-30	Guatemala	CENICAFE provenance and progeny trial	Fredonia (Antioquia)
69	MAR96-09	Guatemala	CENICAFE provenance and progeny trial	Fredonia (Antioquia)
70	061/96	El Salvador	CENICAFE provenance and progeny trial	Fredonia (Antioquia)
71	ABR95-30	Guatemala	CENICAFE provenance and progeny trial	Fredonia (Antioquia)

Table 1. Samples used, codes, country of origin, and type of trial to characterize *Tabebuia rosea* in Colombia.

2.2. Molecular characterization

Molecular characterization was performed using microsatellite markers in two ways:

1. Applying the principle of transferability: markers developed for *T. aurea* by Braga et al. [16] were used in *T. rosea*.

2. Developing specific microsatellites for *T. rosea* from genomic libraries enriched for microsatellite motifs.

In both cases, the same plant material was used as well as the same DNA extraction protocols and electrophoresis and silver nitrate staining procedures; however, amplification conditions varied depending on the group of microsatellite markers. Amplified products were separated by denatured 6% polyacrylamide gel electrophoresis, run in BioRad's Sequi-Gen GT Sequencing Cell vertical electrophoresis chambers. Gels were dyed with silver nitrate following the protocol of Benbouza et al. [17].

Details of the procedure are described below:

2.3. DNA extraction

The QIAGEN Plant DNeasy Mini Kit was used to extract DNA from 20 mg macerated dry leaf tissue following the manufacturer's recommendations. To improve quality, the DNA was purified using the protocol described by Castillo [18]. DNA quality and quantity were verified by agarose gel electrophoresis at 0.8%, dyed with ethidium bromide.

2.4. DNA amplification using microsatellites developed for *Tabebuia aurea*

Based on the principle of transferability, DNA amplification of *T. rosea* was performed using 21 microsatellites developed by Braga et al. [17] for *T. aurea* (Bignoniaceae), which were identified with the prefix 'Tau' (Table 2).

The amplification protocol was carried out in a final volume of 15 µl, with 0.9 µM of each primer, 1U Taq polymerase, 200 µM of each dNTP, 1X reaction buffer (10 mM Tri HCl, 50 mM KCl, 1.5 mM MgCl$_2$) and 10 ng DNA. The amplification profile consisted of 30, 1-minute cycles at 94 °C, 1 minute at melting temperature (°C) (see Table 2), 1 minute at 72 °C, with a final 10-minute extension at 72 °C.

Locus	GenBank accession number	Primer sequence F (5′-3′)	Primer Sequence R (5′-3′)	Annealing temperature (° C)	Number of loci	He[1]	Ho[1]	PIC[1]
Tau07	DQ666983	CCATAAGCTGCATCAACACA	ATCCTAAGATCGGTACTCCA	50	1	0.4259	0.478	0.385
Tau12	DQ666987	CATCATCAAGGTCAAGATCA	CATTCTAGTCTTCCATAAGT	52	1	0.6234	0.491	0.5513
Tau14	DQ666989	GGTAACGGATTGCTGGTTGT	CATTGCGAATGGCCTATGGT	55	1	0.8344	0.721	0.816
Tau17	DQ666992	TGGCCGTGTTGATGTTTATG	TGCCTCACGCTCTATGTGTC	52	1	0.8439	0.701	0.8269
Tau21	DQ666996	CTTTTGGGGGTCTTTGGAAT	TGAAAGAGACAGAGACAAAGATACA	55	1	0.5285	0.290	0.4911
Tau22	DQ666997	TATCTCTCCGCCGTACACCT	CCAATCGAAGAGCCCATTTA	52	2	0.7803	0.576	0.7502
Tau27	DQ667000	GGTAAATCATCTTCCGCTTCC	ACTGCAGAATCGCCTTTTGT	52	1	0.4081	0.508	0.3684
Tau30	DQ667002	TAGTTTAAGGGTGCCGTTGG	CGAACATAAAGAGGCAACCA	55	2	0.4558	0.588	0.3942
Tau31	DQ666982	TCGTGCAGCTTTTGAGTCTG	CTGCAAAACACAAAGCGAAA	52	1	0.6982	0.632	0.6492
TRA101	GU011737	CAAGACACATCCACGTACATAG	CTCACTCCCTTTAGTTTGTCAC	56	2	0.7749	0.568	0.7454
TRA3	GU011814	AGTAATTCCATCCAATCACATC	TGCATCAATCAAGTTGTAAGTC	56	2	0.4154	0.422	0.3715
TRA104	GU011690	CTCCCAAAGCCTTCTTTATATC	GTGGTAGTTGGAGAACATCATC	56	2	0.6185	0.984	0.5431

Locus	GenBank accession number	Primer sequence F (5′-3′)	Primer Sequence R (5′-3′)	Annealing temperature (° C)	Number of loci	He[1]	Ho[1]	PIC[1]
TRB109	GU011820	GCGCTGATGTTCATAATCTGA	CCATTGTTGGCCCTATCTTAT	56	2	0.7929	0.735	0.7606
TRB110	GU011754	GACCCAGGAAATGTTCTCG	AACGGTTGAGGAGCCATC	56	2	0.7308	0.375	0.6912
TRB103	GU011713	GAACGGGAAGACGCAGTC	GGCAGGTGGCAGAAGATC	56	2	0.8288	0.499	0.8070
TRB6	GU011721	TCATTGAGAGGAGCATTATACA	TTCAGTTGCGATGAGACAG	56	1	0.7783	0.590	0.7536
TRB8a	GU011719	GGTGGTGGAACGTCAGAT6AAG	GAGGGAATGCAAACACTTCAC	55	5	0.7686	0.576	0.7471
TRD1	GU011808	CCATCCATCACATCAAGC	GAAAGCAGTTCCCAGTAGTG	55	4	0.4267	0.459	0.3841
TRB104	GU011740	GTTCAATATGCGTCATCAATC	AACGAACTCAGAACTTTCGAC	55	1	0.8103	0.615	0.783
TRC8	GU011704	TTGGCTGACTGATACGATTG	GTGCTGGTCCTGTCCATC	55	1	0.5732	0.353	0.5557
TRA109	GU011728	GGAGAACGGATGTCTGTCAG	GCGTAGGATTTGGTGAAGTG	55	2	0.6998	0.576	0.6479
TRD110	GU011708	TGGATTAGAGAGCATGAGG	GCCATAATGATCCTGCATG	55	1	0.409	0.333	0.3916
TRC103	GU011717	TATTTCGCTCACGCATAAG	GCTTTGTCTCCTATCCAACTC	55	1	0.8433	0.750	0.8232
TRC105	GU011770	AAGCCCAGATTACTGTCTTCC	CGCGTGTGAGACTGTGAC	55	1	0.6885	1.000	0.6348

[1] He: Expected heterozygosity; Ho: Observed heterozygosity; PIC: polymorphic information content.

Table 2. Characteristics of the 24 microsatellite markers used in *Tabebuia rosea*.

2.5. Construction of genomic libraries to design primers for *Tabebuia rosea*

A genomic library enriched for microsatellite motifs was built to obtain a higher number of microsatellites. The microsatellites developed using this methodology was identified with prefix 'TR'. The procedure for building the library is described below:

Colony production. Recombinant plasmids were produced by linking digested fragments of *T. rosea* in the Hind *III* restriction site of plasmid pUC19. Fragments were enriched for microsatellite motifs. Digested products were introduced by electroporation within *Escherichia coli* DH5α races (ElectroMax™, Invitrogen). To isolate colonies for subsequent sequencing, cells were planted on Petri dishes with culture media containing Blue-gal/IPTG/ampicillin-LB agar, X-gal/IPTG/ampicillin-LB agar, or S-gal/IPTG/ampicillin-LB agar.

Clone selection for sequencing. Fragments containing microsatellite sequences were selected using biotinylated oligonucleotides complementary to the AG/CT repeat sequences and then amplified by PCR. DNA inserts were sequenced using the Amersham DYEnamic™ kit, following the manufacturer's instructions, and then run on electrophoresis in an ABI 377 sequencer (Applied BioSystems).

Microsatellite design. Before designing the primers, vector contamination was suppressed in the clone sequences and the percentage of redundancy in the genomic library was determined. PCR primers were designed using the software DesignerPCR, version 1.03 (Research Genetics, Inc). Table 2 presents the SSRs used in this study.

2.6. Amplification of microsatellites designed for *Tabebuia rosea*

Amplification reactions were conducted in a final volume of 10 µl with 0.6 µm of each primer, 200 µm of each dNTP, 1X reaction buffer, 1 U Taq polymerase, 2 mM $MgCl_2$, and 2 ng DNA. The amplification profile was 94 °C, 4 minutes of initial denaturation, 30 cycles, 40 seconds each, at 94 °C, melting temperature (° C) (see Table 2) for 40 seconds, 72 °C for 30 seconds, and a final 4-minute extension at 72 °C.

2.7. Statistical analysis

Statistical analyses were carried out using the program GENAlEX version 6.2 [19]. The analysis included the measurement of genetic variability, genetic diversity, and genetic distance.

3. Results and discussion

3.1. Transferability of Tau microsatellites

Of the 21 microsatellite markers evaluated, 11 presented positive amplification in samples of *T. rosea* and, of these, nine presented polymorphic amplifications (Table 2). The transferability of markers between species of the same genus was possible thanks to the synteny described in plant species and to comparative mapping in plants, such as that carried out by Bonierbale et al. [20] in Solanaceae. Other important synteny studies have been carried out in Gramineae to evaluate maize microsatellites in sugarcane [21] and rice microsatellites in sorghum [22]. Rice and sugarcane microsatellite sequences were used to characterize natural populations of *Guadua angustifolia* [23].

The transferability of microsatellites between species of the same genus, in particular tropical trees, has been addressed by Collevatti-Garcia et al. [8] for the *Caryocar* genus and by Braga et al. [16] for the *Tabebuia* genus. The latter study confirmed the transferability of microsatellite markers—the same used in this study—to other species of the *Tabebuia* genus (*T. ochracea, T. serratifolia, T. roseo-alba, T. impetiginosa*). Braga et al. [16] described the primers

Tau 12, Tau 14, Tau 15, Tau 21, Tau 22, Tau 27, Tau 28, and Tau 31 as the most versatile primers because they presented positive amplification in all the aforementioned species of *Tabebuia*, whereas Tau 13, Tau 17, and Tau 30 presented amplification in at least three of the four *Tabebuia* species. Of the first group described by Braga et al. [16], Tau 12, Tau 14, Tau 21, Tau 22, and Tau 31 amplified samples of *T. rosea* in the present study and, of the second group, Tau 17 and Tau 30.

The nine polymorphic microsatellites used in the present study showed amplifications of one or two loci for a single microsatellite marker (Table 2), although polymorphic amplification was obtained with nine SSRs and 11 loci were observed. Markers presenting two loci were Tau 22 and Tau 30 (Table 2). The size of the products obtained ranged from 150 to 300 bp. These data agreed with those described for microsatellite development.

The distribution of the allele patterns in the three tree populations of *T. rosea* studied indicate that the average number of alleles in the three populations varied between four and five alleles. The number of frequent alleles ranged between three and four, whereas the number of informative alleles ranged between two and three (Table 2). The latter indicates the allelic diversity throughout all loci in each population. The average number of private alleles (whose frequency is less than 5%) is close to 1, with the CENICAFE population presenting the highest number of private alleles.

3.2. Molecular characterization with TR microsatellites

A total of 109 different microsatellites were identified in four of the enriched genomic libraries, which contained between 10,000-15,000 recombinant cells and allowed 74 microsatellite primers to be designed and 24 primers to be synthesized. The percentage of redundancy for this library was 11.4%. The latter were initially tested with the six samples that presented the highest polymorphism in analysis with Tau primers. The seven microsatellite primers presenting polymorphic amplification (Table 2) were then selected. Each of these seven primers presented amplification of two loci, except for the TRB6 primer pair, which presented the amplification of one locus (Table 2).

The results obtained based on the construction of the genomic library can be compared with those obtained by Braga et al. (2007), who obtained 271 positive colonies after constructing enriched libraries in the same genus, from which 31 polymorphic primers were synthesized.

For the most part, the microsatellites developed for *T. rosea* (TR primers) allowed the amplification of two loci per primer pair. Multi-allelic SSRs, such as TRB103 and TRB6, could be differentiated based on the allele frequencies of each locus in each population. With the primer pair TRB6, private alleles occur in the CENICAFE population (alleles 9, 10, and 11). The second locus of TRA3 presents a high frequency (over 80%) of allele 2 in the three populations. This allele could serve as marker for the species. The distribution of allele patterns indicates that the total number of alleles ranges between 4 and 6 and the number of informative alleles between 3 and 4.

Other studies carried out in forest species, such as the one conducted by Li et al. [24], report an average of 5-7 informative alleles for *Quercus aquifolioides*, using six microsatellites, after having obtained 12 alleles on average. These results ratify the phenomenon observed in the present study, where after having an average of 4-6 alleles, the number of informative alleles declined drastically to only 3-4. Braga et al. [16] reported an average of 18.7 alleles for *T. aurea*, which differs from the value found in this study (average of 4-6 alleles). This difference could be attributed to the fact that Braga et al. [16] worked with natural populations, whereas the present study was conducted in selected populations where several alleles could have been lost. This hypothesis gains strength if the number of alleles obtained for the two groups of Tau and TR primers is compared. The results are similar. In addition, with both Tau and TR primers, the highest number of alleles is observed in selected REFOPAL-CONIF and CENI-CAFE populations.

3.3. Molecular characterization with both groups of SSRs (Tau and TR)

The results of the analyses of the 24 loci corresponding to both groups of microsatellites (Tau and TR) are analyzed and discussed below.

3.4. Measures of genetic variability (private alleles)

The evaluation of the presence of private alleles indicated a marked prevalence of this type of alleles in CENICAFE provenance and progeny populations and in REFOPAL-CONIF clonal orchards, indicating the need to preserve these orchards to perform directed crosses.

Lombardi [25] found a total of 85 alleles, with an average of 10.6 alleles per locus, when samples of *T. roseo-alba* were amplified with markers developed by Braga et al. [16]. Feres et al. [26] found, in turn, an average of 4.4 alleles per locus in *T. roseo-alba*, values very similar to those found in the present study. Averages of 13.5 alleles have been reported for other Bignoniaceae such as *Jacaranda copaia* [27].

3.5. Measures of genetic diversity

The genetic diversity parameters found when analyzing the 24 loci (Tau and TR primers) helped identify populations with low consanguinity, multi-allelic markers, and expected heterozygosities equal to observed heterozygosities. The mean expected heterozygosity is similar to the mean observed heterozygosity (Table 3). These values also indicate that the frequency of several alleles in the population is clearly concentrated in a few alleles. Values higher than 0.8 have been obtained in multi-allelic microsatellites in tropical forest species, those presenting between 9-10 alleles [8]. This finding differs from the results obtained in this study, where the average number of alleles for all loci and populations ranged between 4 and 5 with the microsatellites used. Findings suggest that these populations, although selected for genetic improvement, are balanced. Studies in *T. aurea* with microsatellites specifically developed for the species revealed an average of 26 alleles per locus and an observed heterozygosity of 0.587, a value similar to that reported for *T. rosea*. In addition, the probability of genetic identity was high (1.03×10^{-37}) and the probability of exclusion was 0.9889 [16].

Population	Na* (mean)	Ne* (mean)	He* (mean)	Ho* (mean)
REFOCOSTA-CONIF (clonal orchards)	5.333	3.374	0.630	0.608
REFOCOSTA-CONIF (progenies)	4.250	2.937	0.598	0.635
CENICAFE (provenance and progeny trials)	4.875	2.907	0.608	0.502
All populations/all loci	4.819	3.072	0.612	0.582

* Na=Number of alleles with frequencies higher than or equal to 5%; Ne=Number of informative alleles; He=Expected heterozygosity; Ho=Observed heterozygosity

Table 3. Genetic diversity in three populations for forest improvement programs of *Tabebuia rosea* and two groups of microsatellites (Tau and TR).

Other forest species of interest, such as pine species *Pinus elliottii* var. elliottii and *P. caribaea* var. hondurensis, yielded heterozygosity values of 0.52+/-0.2. Shepherd et al. [28] attribute these values to the use of transpecific microsatellites in pine species, a situation which is replicated in this study by using *T. aurea* microsatellites in *T. rosea*.

Lombardi [25], who also worked with Tau primers, reported a mean observed heterozygosity of 0.271, compared with a heterozygosity of 0.746. Feres et al. [26] attribute the differences between heterozygosities to the presence of null alleles that mask heterozygote individuals under the presence of one band. Null alleles occur more frequently in transferred primers than in primers designed specifically for a given species.

The analysis of molecular analysis (AMOVA) revealed that the highest percentage of variation could be attributed to differences between individuals (71%), followed by the variation between populations (25%) and variation within populations (4%). Miller and Schaal [29] analyzed the structure of natural and cultivated populations of a forest species presenting an edible fruit (*Spondias purpurea*). The percentage of variation between natural populations was 35.65% and, within populations, 64.35%, whereas the corresponding values in cultivated materials were 30.19% and 69.81%.

3.6. Genetic distance measurements

The genetic distance between populations indicates that the populations that are closest genetically (0.114) are those of REFOPAL–CONIF (both clonal orchards and progenies). This value was expected because of the higher inter-relationship of the trees of this group. The CENICAFE population, on the other hand, is distant from populations on Colombia's Atlantic Coast.

Principal coordinates analysis, based on the genetic distances between individuals, revealed greater proximity between the REFOPAL-CONIF populations—clonal orchards and progenies—and a greater distance regarding the CENICAFE provenance and progeny trial (Figure 2). The genetic distance between the populations of the CENICAFE trials and the REFOPAL–CONIF trials (both clonal orchards and progenies) is evident. CENICAFE trials present greater variability than REFOPAL–CONIF trials and this could be attributed to the introduction of Central American materials to CENICAFE's genetic breeding program. Figure 2 endorses the idea that both breeding programs are complementary, and that the exchange of material and information can prove advantageous for both entities. If plans are to begin controlled pollinations, then it would be practical to cross materials from both programs. Likewise, tree clones of both programs should be established in open-pollinated orchards.

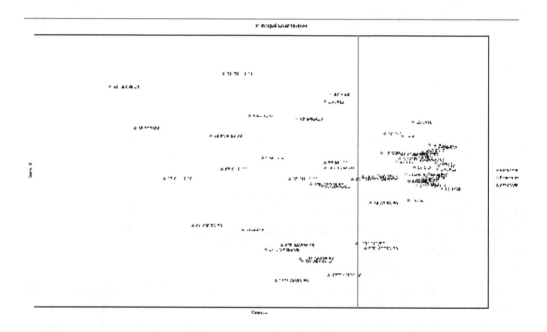

Figure 2. Principal coordinates analysis of samples of *Tabebuia rosea*.

Probability of identity: Given the need to identify each individual and assign each a probability of unequivocal identity, the study showed that this identity was established by combining eight different loci. More multi-allelic markers, with higher indexes of heterozygosity, are preferred. The optimal combination to characterize accessions of Vitis vinifera has also been reported to be eight primers [30].

Kirst et al. [31] determined the probability of identity of *Eucalyptus grandis* using microsatellites. The probability of finding identical genotypes in a population was found to be 2.3 x 10^{-9} using six EMBRA microsatellites developed in Brazil. This probability is considered practically null. Similar results were obtained with an improved population of *Eucalyptus dunnii*, when 46 individuals from a seed orchard were identified using four microsatellites [32].

4. Conclusions

This study verifies the transferability of microsatellites existing in the *Tabebuia* genus and sets forth the principle of synteny as an economically viable alternative for other *Tabebuia* species for which information on the development of molecular markers is not available. The ample use of the microsatellites developed for *T. aurea* (Tau microsatellites) in several species of *Tabebuia* proves useful for the exchange of information between genetic breeding and conservation groups of *Tabebuia* species at the national and international levels.

Both groups of microsatellites amplified 24 polymorphic loci that could be useful in genotyping individuals of *T. rosea*. The measurement of genetic variability yielded an average of 4-5 alleles and between 3-4 frequent alleles. Expected heterozygosities equal to observed heterozygosities indicate that there is equilibrium within the populations, despite dealing with selected materials. The low indices of consanguinity of the three populations originating from improvement programs confirm the allogamous behavior of the species and, with the previous results, the hypothesis that the flower is self-incompatible gains strength. Meanwhile, private alleles were more frequent in the CENICAFE provenance and progeny trials and the REFOPAL-CONIF clonal orchards. The introduction of Central American materials influenced the high variability of CENICAFE materials. As a result, both selection processes are non-exclusive; on the contrary, they are complementary and the exchange of plant materials and information will prove beneficial to both.

Acknowledgements

The authors express their sincere thanks to Dr. Marco Cristancho (CENICAFE) for the technical revision of the present manuscript, to the Ministry of Agriculture and Rural Development of the Republic of Colombia, to the National Federation of Coffee Growers of Colombia, to the Government of the Department of Risaralda, and to the Technical University of Pereira for funding this research; to REFOPAL and CONIF for providing plant materials and morphological information on the evaluated materials. Our sincere thanks also go to Juliana Arias Villegas for planning and managing the project and to the staff of the Plant Biotechnology Lab of the Technical University of Pereira for their valuable contributions throughout the present study.

Author details

Ana María López[1*], Marta Leonor Marulanda[1] and Carlos Mario Ospina[2]

*Address all correspondence to: alopez@utp.edu.co

1 Universidad Tecnológica de Pereira, Plant Biotechnology Laboratory, Colombia

2 Cenicafé, National Federation of Coffee Growers of Colombia, Colombia

References

[1] Espinal CF., Martínez H., and Gonzalez, E. Características y estructura del sector forestal-madera-muebles en Colombia: Una mirada global de su estructura y dinámica 1991-2005. Ministerio de Agricultura y Desarrollo Rural, Observatorio Agrocadenas Colombia. Documento de trabajo No. 95. 2005.

[2] Burley J. Forest genetics for sustainable forest management. In: Plantation technology in tropical forest science. Edited by K. Suzuki, K. Ishii, S. Sakurai, and S. Sasaki. Springer. Tokyo. pp. 336. 1996.

[3] Intergovernmental Panel on Climate Change (IPCC). Land use, land use change and forestry. A special report of the Intergovernmental Panel on Climate Change (IPCC), Cambridge University, United States of America. 377 p. 2000.

[4] Dale G., and Chaparro J. Integration of molecular markers into tree breeding and improvement programs. In Tree improvement for sustainable tropical forestry. Edited by M.J. Dieters, A.C. Matheson, D.G. Nikles, C.E. Harwood, and S.M. Walker. QFRI–IUFRO, Australia. pp. 472-477. 1996.

[5] Wickneswari R., Norwati M., Lee SL., Lee CT., Yeang HY., Lokmal N., and Rasip AGA. Practical uses of molecular markers in tropical tree improvement. In Proceedings of the QFRI-UIFRO Conference of Tree Improvement for Sustainable Tropical Forestry. Edited by M.J. Dieters, A.C. Matheson, D.G. Nikles, C.E. Harwood, and S.M. Walker. Queensland, Australia. pp. 506-511. 1996.

[6] Ruane J., and Sonnino A. Marker-assisted selection as a tool for genetic improvement of crops, livestock, forestry and fish in developing countries: an overview of the issues. In Marker-assisted selection: Current status and future perspectives in crops, livestock, forestry and fish. Edited by E.P. Guimarães, J. Ruane, B.D. Scherf, A. Sonnino, and J.D. Dargie. Food and Agriculture Organization of the United Nations, Rome. pp. 3-13. 2007.

[7] Grattapaglia D. Marker-assisted selection in *Eucalyptus*. In Marker-assisted selection: Current status and future perspectives in crops, livestock, forestry and fish. Edited by E.P. Guimarães, J. Ruane, B.D. Scherf, A. Sonnino, and J.D. Dargie. Food and Agriculture Organization of the United Nations, Rome. pp. 251-282. 2007.

[8] Collevatti-García R., Brondani V., and Grattapaglia D. Development and characterization of microsatellite markers for genetic analysis of a Brazilian endangered tree species *Caryocar brasiliense*. Heredity 1999; 83: 748-756.

[9] Brondani, C., Rangel, P., Oliveira, T., Pereira, R., and Brondani, V. Transferability of microsatellite and sequence tagged site markers in *Oryza* species. Hereditas 2003; 138(3): 187-192.

[10] Ciampi A.Y., Brondani R.P.V., and Grattapaglia D. Desenvolvimento de marcadores microssatélites para *Copaifera langsdorffii* Desf. (Copaíba)-Leguminosae-Caesalpinioi-

deae e otimização de sistemas fluorescentes de genotipagem multiloco. Boletim de Pesquisa 2000 (16). Empresa Brasileira de Pesquisa Agropecuária (Embrapa), and Centro Nacional de Pesquisa de Recursos Genéticos (Cenargen), Brasília (Distrito Federal). 2000.

[11] Ferreira-Ramos R., Laborda P., Oliveira Santos M., Mayor M., Mestriner M., Souza A., and Alzate-Marin A.L. Genetic analysis of forest species *Eugenia uniflora* L. through newly developed SSR markers. Conserv. Genet. 2006; 9(5): 1281-1285.

[12] Guidugli M., De Campos T., Barbosa de Sousa A.C., Massimino Feres J., Sebbenn A.M., Mestriner M.A., Betioli Contel E.P., and Alzate-Marin A.L. Development and characterization of 15 microsatellite loci for *Cariniana estrellensis* and transferability to *Cariniana legalis*, two endangered tropical tree species. Conserv. Genet. 2009; 10(4): 1001-1004.

[13] Centro Nacional de Investigaciones de Café (CENICAFE), and Federación Nacional de Cafeteros de Colombia (FEDECAFE). Ensayo de procedencias y progenies para dos especies forestales tropicales de alto valor comercial: Resultados marzo 1996-febrero 2000. Special collaborative agreement between FEDERACAFE-Ministry of the Environment of Colombia for forestry research with native species. 2000.

[14] Hoyos J. Flora emblemática de Venezuela. Armitano, Caracas, Venezuela. 213 p. 1985.

[15] Blanco-Mavares C.M. 25 Árboles de Caracas. FUNDARTE, Alcaldía de Caracas, Caracas, Venezuela. 72 p. 1995.

[16] Braga A., Reis A.M.M., Leoi L., Pereira R., and Collevatti R. Development and characterization of microsatellite markers for the tropical tree species *Tabebuia aurea* (Bignoniaceae). Mol. Ecol. Notes 2007; 7: 53-56.

[17] Benbouza H., Jacquemin J.M., Baudoin J.P., and Mergeai G. Optimization of a reliable, fast, cheap and sensitive silver staining method to detect SSR markers in polyacrylamide gels. Biotechnologie, Agronomie, Société et Environnement 2006; 10(2): 77-81.

[18] Castillo N.R. Fingerprinting and genetic stability of *Rubus* using molecular markers. M.Sc. thesis, Oregon State University, Oregon. 244 p. 2006.

[19] Peakall R., and Smouse P.E. GENALEX 6: genetic analysis in Excel. Population genetic software for teaching and research. Mol. Ecol. Notes 2006; 6: 288-295.

[20] Bonierbale M.W., Plaisted R.L., and Tanksley S.D. RFLP maps based on a common set of clones reveal models of chromosomal evolution in potato and tomato. Genetics 1998; 120: 1095-1103.

[21] Selvi A., Nair N.V., Balasundaram N., and Mohapatra T. Evaluation of maize microsatellite markers for genetic diversity analysis and fingerprinting in sugarcane. Genome 2003; 46: 394-403.

[22] Ishii T., and McCouch S.R. Microsatellites and microsynteny in the chloroplast genomes of *Oryza* and eight other Gramineae species. Theor. Appl. Genet. 2000; 100:
1257-1266.

[23] Marulanda M.L., López A.M., and Claros J.L. Analyzing the genetic diversity of *Guadua* spp. in Colombia using rice and sugarcane microsatellites. Crop Breed. Appl. Biotechnol. 2007; 7(1): 43-48.

[24] Li Ch., Zhang X., Liu X., Luukkanen O., and Berninger F. Leaf morphological and
physiological responses of *Quercus aquifolioides* along an altitudinal gradient. Silva
Fennica 2006; 40(1): 5-13.

[25] Lombardi M. Aplicação de marcadores microssatellites na caracterização de recursos
genéticos de *Tabebuia roseo–alba* conservados ex situ no Banco de Germoplasma da
Floresta da USP de Ribeirão Prieto. M.Sc. thesis, Departamento de Biologia, Universidad de Sao Paulo, Ribeirão Prieto, Brazil. 126 p. 2008.

[26] Feres J., Martinez M., Martinez C., Mestriner M., and Alzate-Marin A.L. Transferability and characterization of nine microsatellite markers for the tropical tree species
Tabebuia roseo-alba. Mol. Ecol. Resour. 2008; 9: 434-437.

[27] Jones F., and Hubbell S.P. Isolation and characterization of microsatellite loci in the
tropical tree *Jacaranda copaia* (Bignoniaceae). Mol. Ecol. Notes 2003; 3(3): 403-405.

[28] Shepherd M., Cross M., Maguire T.L., Dieters M.J., Williams C.G., and Henry R.J.
Transpecific microsatellites for hard pines. Theor. Appl. Genet. 2002; 104: 819-827.

[29] Miller A.J., and Schaal B.A. Domestication and the distribution of genetic variation in
wild and cultivated populations of the Mesoamerican fruit tree, *Spondias purpurea* L.
(Anacardiaceae). Mol. Ecol. 2006; 15: 1467-1480.

[30] Tessier C., David J., This P., Boursiquot J.M., and Charrier A. Optimization of the
choice of molecular markers for varietal identification of *Vitis vinifera* L. Theor. Appl.
Genet. 1999; 98(1): 171-177.

[31] Kirst M., Cordeiro C.M., Rezende G.D.S.P., and Grattapaglia D. Power of microsatellite markers for fingerprinting and parentage analysis in *Eucalyptus grandis* breeding
populations. J. Hered. 2005; 96(2): 161-166.

[32] Marcucci Poltri S.N., Zelener N., Rodriguez Traverso J., Gelid P., and Hopp H.E. Selection of a seed orchard of *Eucalyptus dunnii* based on genetic diversity criteria calculated using molecular markers. Tree Physiol. 2003; 23: 625-632.

9

Use of Organelle Markers to Study Genetic Diversity in Soybean

Lidia Skuza, Ewa Filip and Izabela Szućko

1. Introduction

Soybean is the most important crop provider of proteins and oil used in animal nutrition and for human consumption. Plant breeders continue to release improved cultivars with enhanced yield, disease resistance, and quality traits. It is also the most planted genetically modified crop. The narrow genetic base of current soybean cultivars may lack sufficient allelic diversity to counteract vulnerability to shifts in environmental variables. An investigation of genetic relatedness at a broad level may provide important information about the historical relationship among different genotypes. Such types of study are possible thanks to different markers application, based on variation of organelle DNA (mtDNA or cpDNA).

2. Mitochondrial genome

2.1. Genomes as markers

Typically, all sufficiently variable DNA regions can be used in genetic studies of populations and in interspecific studies. Because of in seed plants chloroplasts and mitochondria are mainly inherited uniparentally, organelle genomes are often used because they carry more information than nuclear markers, which are inherited biparentally. The main benefit is that there is only one allele per cell and per organism, and, consequently, no recombination between two alleles can occur. With different dispersal distances, genomes inherited biparentally, maternally and paternally, also reveal significant differences in their genetic variability among populations. In particular, maternally inherited markers show diversity within a population much better [1].

In gymnosperms the situation is somewhat different. Here, chloroplasts are inherited main-ly paternally and are therefore transmitted through pollen and seeds, whereas mitochondria are largely inherited maternally and are therefore transmitted only by seeds [2]. Since pollen is distributed at far greater distances than seeds [3], mitochondrial markers show a greater population diversity than chloroplast markers and therefore serve as important tools in con-ducting genetic studies of gymnosperms [4]. Mitochondrial markers are also sometimes used in conjunction with cpDNA markers [5].

Mitochondrial regions used in interspecific studies of plants, mainly gymnosperms, in-clude, for example, introns of the NADH dehydrogenase gene *nad1* [4, 5, 6], the *nad7* in-tron 1 [7], the *nad5* intron 4 [3] and an internally transcribed spacer (ITS) of mitochondrial ribosomal DNA [8, 9].

In addition to the aforementioned organelle markers, microsatellite markers [10, 11] and simple sequence repeats (SSR) are often used in population biology, and sometimes also in phylogeographic studies. Microsatellites are much less common in plants than in animals [12]. However, they are present in both the nuclear genome and the organelle genome. Mi-crosatellites may reveal a high variability, which may be useful in genetic studies of popula-tions, whereas other sequences or methods such as fingerprinting do not detect mutations sufficiently [9,10,13]. Inherited only uniparentally, organelle markers have a certain quality in phylogeographic analyses. Since they are haploid, the effective population size should be reduced after the analysis using these markers as compared to those in which nuclear mark-ers are used [1, 14]. Smaller effective populations sizes should bring about faster turnover rates for newly evolving genotypes, resulting in a clearer picture of past migration history than those obtained using nuclear markers [15-17].

Initially, it was mainly in phylogeographic studies of animal species that mitochondrial markers were used [18]. These studies have provided some interesting data on the begin-nings and the evolutionary history of human population [19]. In contrast to studies of ani-mals, using mitochondrial markers in studies of plants, especially angiosperms, is limited [20]. Presently, cpDNA markers are most commonly used in phylogeographic studies of an-giosperms, whereas mitochondrial markers are prevalent in studies of gymnosperms.

2.2. Plant mitochondrial DNA

Mitochondrial genomes of higher plants (208-2000 kbp) are much larger than those of verte-brates (16-17 kbp) or fungi (25-80 kbp) [21, 22]. In addition, there are clear differences in size and organization of mitochondrial genomes between different species of plants. Intramolec-ular recombination in mitochondria leads to complex reorganizations of genomes, and, in consequence, to alternating arrangement of genes, even in individual plants, and the occur-rence of duplications and deletions are common [23]. In addition, the nucleotide substitution rate in plant mitochondria is rather low [24], causing only minor differences within certain loci between individuals or even species. Extensively characterized circular animal mito-chondrial genomes are highly conservative within a given species; they do not contain in-trons and have a very limited number of intergenic sequences [25]. Plant mitochondrial DNA (mtDNA) contains introns in multiple genes and several additional genes undergoing expression when compared to animal mitochondria, but most of the additional sequences in plants are not expressed and they do not seem to be esssentials [26]. The completely sequenced mitochondrial genomes are available for several higher plants, including *Arabidopsis thaliana* [27] or *Marchantia polymorpha* [28].

Restriction maps of nearly all plant mitochondrial genomes provide for the occurrence of the master circle with circular subgenomic molecules that arise after recombination among large direct repeats (> 1 kbp) [21, 29-36], which are present in most mitochondrial genomes of higher plants. However, such molecules, whose sizes can be predicted, are very rare or very difficult to observe. It can be explained by the fact that plant mitochondrial genomes are circularly permuted as in the phage T4 [37, 38]. Oldenburg and Bendich reported that mostly linear molecules in *Marchantia* mtDNA are circularly permuted with random ends [39]. It shows that plant mtDNA replication occurs similarly to the mechanism of recombination in the T4 [38].

Many reports that have appeared in recent years indicate that mitochondrial genome of yeasts and of higher plants exist mainly as linear and branched DNA molecules with variable size which is much smaller than the predicted size of the genomes [39-44]. Using pulsed field gel electrophoresis (PFGE) of in-gel lysed mitochondria from different species revealed that only about 6-12% of the molecules are circular [41, 44]. The observed branched molecules are very similar to the molecules seen in yeast in the intermediate stages of recombination of mtDNA [45] or the phage T4 DNA replication [37, 38].

In all but one known case (*Brassica hirta*) [46], plant mitochondrial genomes contain repeat recombinations. These sequences, ranging in length from several hundred to several thousand nucleotides (nt) exist at two different loci in the master circle, yet in four mtDNA sequence configurations [47]. These four configurations correspond to the reciprocal exchange of sequences 5' and 3' surrounding the repeat in the master circle, which suggests that the repeat mediates homologous recombination. Depending on the number and orientation of repeats, the master circle is a more or less complex set of subgenomic molecules [48].

Maternally inherited mutations, which are associated with mitochondria in higher plants, most often occur as a result of intra- and intergenic recombination. This happens in most cases of cytoplasmic male sterility (cms) [41, 49-51], in *chm*-induced mutation in *Arabidopsis* [52] and in non-chromosomal stripe mutations in maize [53]. In this way, it is assumed that the recombination activity explains the complexity of the variations detected in the mitochondrial genomes of higher plants.

2.3. Mitochondrial genome of soybean

The size of soybean mtDNA has been estimated to be approximately 400 kb [54-56]. Spherical molecules have also been observed by electron microscopy [55, 57].

Repeated sequences 9, 23 and 299 bp have been characterized in soybean mitochondria [58, 59]. Also, numerous reorganizations of genome sequences have been characterized among different cultivars of soybean. It has been demonstrated that they occur through homologous recombination produced by these repeat sequences [58, 60, 61], or through short elements that are part of 4.9kb PstI fragment of soybean mtDNA [62]. The 299 bp repeat sequence has been found in several copies of mtDNA of soybean and in several other higher plants, suggesting that this repeated sequence may represent a hot spot for recombination of mtDNA in many plant species [59, 62]. Previous results suggested that active homologous recombinations of mtDNA are present in at least some species of plants. Recently (2007) amitochondrial-targeted homolog of the *Escherichia coli recA* gene in *A. thaliana* has been identified [63]. However, the data on recombnation activity in plant mitochondria is still missing. The first data on such an activity in soybean was obtained in 2006 [64]. This discovery is supported by an analysis of mtDNA of soybean using electron microscopy and 2D-electrophoresis. The results suggest that only a small portion of mtDNA molecules undergoes recombination at any given time. Therefore the question is whether this recombination is essential to the functioning of mitochondria and to plant growth.

The repeated sequences of the *atp6*, *atp9* and *coxII* genes have been also characterized, but their recombination activity has not been analysed [65].

The first data for the restriction map of soybean mtDNA were obtained from the analysis of loci of the *atp4* gene [48]. In the vicinity of this gene two repeated sequences that show characteristics of recombination repeats have been found [47, 48]. Active recombination repeats were also identified in circular molecules smaller than 400 kb [55, 66]. These observations suggest that soybean mtDNA has multipartite structure that is similar to other plant mitochondrial genomes containing recombination repeats.

In the mitochondrial genome of cultivar Williams 82, recombinantly active repeats 1 kb and 2 kb have been described [48]. In a different repeat of 10 kb, surrounding both 1 kb and 2 kb repeats, two breakpoints have been identified. This recombination of smaller and larger repeats probably leads to the complex structure of genomes.

The analysis of restriction fragment length polymorphism (RFLP) of mtDNA seems to be a useful method in studying phylogenetic relationships within species.

Grabau et al. (1992) analyzed the genomes of 138 soybean cultivars [60]. Using 2.3 kb HindIII mtDNA probe from Williams 82 soybean cultivar revealed restriction fragment length polymorphisms (RFLPs), which allowed for the division of many soybean cultivars into four cytoplasmic groups: Bedford, Arksoy, Lincoln and soja-forage.

Subsequent analyses showed variations within, and adjacent to, the 4.8 kb repeats. Bedford cytoplasm turned out to be the only one that contains copies of the repeat in four different genomic environments, which indicates its recombination activity [61]. Lincoln and Arksoy cytoplasms contain two copies of the repeat and a unique fragment that appear to result from rare recombination events outside, but near, the repeat. In contrast, forage-soja cytoplasm contains no complete repeat, but it contains a unique truncated version of the repeat [61]. Sequence analysis revealed that truncating is caused by the recombination with a repeat of 9 bp CCCCTCCCC. The structural reorganization that occurred in the region around 4.8 kb repeat may provide a way to analyze the relationships between species and evolution within the soybean subgenus.

In order to determine the sources of cytoplasmic variability, Hanlon and Grabau (1995) studied the old cultivars of soybeans with the same 2.3-kb *Hind*III fragment and with a mtDNA fragment containing the *atp6* gene [62]. They showed that mtDNA RFLP analysis with these probes is useful for the classification of mitochondrial genomes of soybean. Grabau and Davies (1992) made a general classification of wild soybean using the 2.3-kb *Hind*III as a probe [68].

Mt type	Probe	*coxI*			*coxII*			*atp6*		Reference
	Enzyme	*Hind*III	*Bam*HI	*Eco*RI	*Hind*III	*Bam*HI	*Eco*RI	*Bam*HI	*Eco*RI	
Ic					1,6	5,8		5,0		[69]
Id					1,6	5,8		5,0;6,0 ;12,0		[69]
Ie					1,6	5,8		5,0; 12,0		[69]
Ik					1,6	5,8		5,0; 5,4; 5,8		[69]
IIg					1,3	7,0		1,0; 2,6		[69]
IIIa					1,2	8,5		2,4; 5,0		[69]
IIIb					1,2	8,5		2,9; 5,0		[69]
IIId					1,2	8,5		5,0;6,0; 12,0		[69]
IVa					3,5	8,1		2,4; 5,0		[69]
IVb					3,5	8,1		2,9; 5,0		[69]
IVc					3,5	8,1		5,0		[69]
IVf					3,5	8,1		2,4; 3,5; 5,0		[69]
IVh					3,5	8,1		2,6; 2,9		[69]
IVi					3,5	8,1		5,2; 12,0		[69]
Va					5,8	8,1		2,4; 5,0		[69]
V'j					5,8	15,0		5,0; 6,0		[69]
VIg					1,7	5,8		1,0; 2,6		[69]
VIIg					8,5	15,0		1,0; 2,6		[69]
mtI					1,6	5,8				[69]
mtII					1,3	7,0				[69]
mtIII					1,2	8,1				[69]
mtIV					3,5					[69]
mtV					5,8					[69]
mt-a								2,4; 5,0		[87]
mt-b								2,9; 5,0		[87]
mt-c								5,0		[87]
mt-d								5,0; 6,0; 12,0		[87]
mt-e								5,0; 12,0		[87]

Mt type	Probe Enzyme	coxI HindIII	BamHI	EcoRI	coxII HindIII	BamHI	EcoRI	atp6 BamHI	EcoRI	Reference
mt-f								2,4; 3,5; 5,0		[87]
mt-g								1,0; 2,6		[87]
mt-h								2,6; 2,9		[87]
mt-m								2,9		[87]
mt-n								12,0		[87]
Ic		5,6	0,8; 2,5; 5,0	10,5	1,6	5,8	1,9	5,0	8,2; 12,0	[58]
Id		5,6	0,8; 2,5; 5,0	10,5	1,6	5,8	1,9	5,0; 6,0; 12,0	2,8; 6,0; 12,0	[58]
Ie		5,6	0,8; 2,5; 5,0	10,5	1,6	5,8	1,9	5,0; 12,0	2,8; 6,0; 12,0	[58]
Ik		5,6	0,8; 2,5; 5,0	10,5	1,6	5,8	1,9	5,0; 5,4; 5,8	2,8; 6,0; 12,0	[58]
IIg		8,5	0,8; 2,5; 5,0	9,0	1,3	7,0	4,8	1,0; 2,6	2,8; 3,0; 9,5	[58]
IIIb		5,6	0,8; 2,5; 5,0	10,5	1,2	8,5	6,2; 6,5	2,9; 5,0	6,0; 8,2; 12,0	[58]
IIId		5,6	0,8; 2,5; 5,0	10,5	1,2	8,5	6,2; 6,5	5,0; 6,0; 12,0	3,2; 6,2; 12,0	[58]
Iva		5,6	0,8; 2,5; 5,0	10,5	3,5	8,1	5,0	2,4; 5,0	3,0; 6,0; 12,0	[58]
IVb		5,6	0,8; 2,5; 5,0	10,5	3,5	5,8	5,0	2,9; 5,0	6,0; 8,2; 12,0	[58]
IVc		5,6	0,8; 2,5; 5,0	10,5	3,5	5,8	5,0	5,0	8,2; 12,0	[58]
IVf		5,6	0,8; 2,5; 5,0	10,5	3,5	5,8	5,0	2,4; 3,5; 5,0	3,2; 6,2; 12,0	[58]
IVh		5,6	0,8; 2,5; 5,0	10,5	3,5	5,8	5,0	2,6; 2,9	3,2; 6,2; 12,0	[58]
IVi		5,6	0,8; 2,5; 5,0	10,5	3,5	5,8	5,0	5,2; 12,0	3,2; 6,2; 12,0	[58]
Va		5,6	0,8; 2,5; 5,0	10,5	5,8	5,8	12,0	2,4; 5,0	3,0; 6,0; 12,0	[58]
Vb		5,6	0,8; 2,5; 5,0	10,5	5,8	5,8	12,0	2,9; 5,0	6,0; 8,2; 12,0	[58]
Vc		5,6	0,8; 2,5; 5,0	10,5	5,8	5,8	12,0	5,0	8,2; 12,0	[58]
V'j		5,6	0,8; 2,5; 5,0	10,5	5,8	15,0	1,6	5,0; 6,0	2,8; 6,0; 12,0	[58]

Mt type	Probe	coxI			coxII			atp6		Reference
	Enzyme	*Hind*III	*Bam*HI	*Eco*RI	*Hind*III	*Bam*HI	*Eco*RI	*Bam*HI	*Eco*RI	
VIg		5,6; 8,5	0,8; 2,5; 5,2	5,0;9,0; 10,5	1,7	5,8	4,5	1,0; 2,6	2,8; 3,0; 4,3; 9,5; 12,0	[58]
VIIg		8,5	0,8; 5,0; 5,2	9,0	8,5	15,0	1,6	1,0; 2,6	2,8; 3,0; 9,5	[58]
VIIIc		5,6	0,8; 2,5; 5,0	10,5	8,5; 10,0	11,0; 15,0	1,6	5,0	8,2; 12,0	[58]
Combined chloroplast and mitochondrial genome type										
cpI +mtIIIb					1,2	8,5		2,9; 5,0		[89]
cpI +mtIVb					3,5	8,1		2,9; 5,0		[89]
cpI+mtIVc					3,5	8,1		5,0		[89]
cpII +mtIVb					3,5	8,1		2,9; 5,0		[89]
cpII +mtIVc					3,5	8,1		5,0		[89]
cpIII+mtIe					1,6	5,8		5,0; 12,0		[89]
cpIII +mtIVa					3,5	8,1		2,4; 5,0		[89]
cpIII +mtVIIIc					8,5; 10,0	11,0; 15,0		5,0		[89]

Table 1. Classification of mitochondrial genome types based on RFLPs using coxI, *coxII* and *atp6* as probes. Sizes of hybridization signals (kb) are shown.

In their research Tozuka et al. (1998) used two fragments of mtDNA as probes: the 0.7-kb *Hind*III-*Nco*I fragment containing the *coxII* (the gene encoding the mitochondrial cytochrome oxidase subunit II) of wild soybean and the 0.66-kb *Sty*I fragment containing the *atp6* (the gene encoding the mitochondrial ATPase subunit 6) from *Oenothera* [69, 70] (Table 1).

Based on the RFLPs detected in gel-blot analysis with the *coxII* and *atp6* probes, the harvested plants were divided into 18 groups. Five mtDNA types were described in 94% of the surveyed plants. The geographical distribution of mtDNA types revealed that in many regions soybean growing wild in Japan consisted of a mixture of plants with different types of mtDNA, sometimes even within a single location. Some of these mtDNA types have shown marked geographic clines among the regions. In addition, some wild soybeans had mtDNA types that were identical to those described in cultivated soybeans. These results suggest that mtDNA analysis could resolve maternal origin among of the genus *Glycine* subgenus *Soja* [69].

Kanazawa et al. (1998) gathered 1097 *G. soja* plants from all over Japan and analyzed their RFLP of mitochondrial DNA (mtDNA) using five probes (*coxI, coxII, atp6, atp9, atp1=atpA*) [58] (Table 1). 20 different types of mitochondrial genomes labeled as combinations of types I to VII and types from a to k were identified and characterized in this study. Nearly all the mtDNA types described for soybean cultivars also occurred in wild soybean.

The mitochondrial *atpA* gene was also analysed [48]. It was shown that in soybean this gene has a sequence in 90-97% identical with mitochondrial genes of other plants [71-81]. Sequence similarity is limited to the *atpA* coding region. An intriguing feature of the *atpA* open reading frame of soybean is an 642 nt overlap in the putative translation termination site onto an unidentified open reading frame of the *orf214*. The ends of the open reading frame contain four tandems of UGA codon that covers four tandems of AUG codon that initiates an unidentified *orf214* frame. The *atpA-orf 214* region was found in soybean mtDNA in multiple sequence contexts. This can be attributed to the presence of two recombination repeats.

The open reading frame shares 79% of nucleotide identity with the *orf214* and is located in the same *atpA* locus position as in common bean *orf209* [82]. Since such organization is a repeat of overlapping the *atpB* and *atpE* reading frames in several chloroplast genes [83, 84], the probability that the *orf214* codes a different ATPase subunit cannot be evaluated because small ATPase subunits are poorly conserved [85].

So for a total of 26 mtDNA haplotypes of wild soybeans have been identified based on RFLP with probes from two mitochondrial genes: *cox2* and *atp6* [69, 86] (Table 1). The three most common haplotypes (Id, IVa and Va) are present in 43 populations. The distribution of mtDNA haplotypes varies among opulations [87]. Recently Shimamoto (2001) analyzed the genetic polymorphisms of mitochondrial genes subgenus *Soja* originating from China and Japan [88] (Table 1). As a result of these studies, 6 types of mitochondrial genomes were distinguished.

3. Chloroplast genome

As the result of the extensive research conducted in the past two decades, cpDNA analysis brought about fundamental changes to the systematics of plants. The chloroplast genome is ideal for phylogenetic analyses of plants for several reasons. First, it occurs abundantly in plant cells and is taxonomically ubiquitous. And since it is well researched, it can be easily tested in the laboratory conditions and analyzed in comparative programs. Moreover, it often contains marker structural features cladistically useful, and, above all, it exhibits moderate or low rate of nucleotide substitution [89]. In regard to the mitochondrial genome, and also to cpDNA, researchers use in their studies two distinct phylogenetic approaches [90], namely taxonomic checking of specific traits features of molecular cpDNA and sequencing of specific genes or regions.

3.1. Chloroplast genome of soybean

In estimating the phylogeny of plants belonging to *Glycine*, particular attention was paid to unusual and specific features of cpDNA. In the course of many studies on the variability of chloroplast genome, a breakthrough came in 1993, with a study on assessing phylogeny of seed plants. The study used a huge database of the nucleotide sequences of the *rbcL* gene [91], encoding the ribulose-1,5-bisphosphate carboxylase, large subunit. The accumulation of a number of comparative data on this chloroplast gene made it a frequent object of research. This is due to the fact that this gene's locus is large (> 1400 bp), and provides many phylogenetically informative traits. The rate of the *rbcL* evolution proved to be appropriate for assessing issues related to phylogeny of plants, especially on the medium and high taxonomic levels. Over the years other sequences from other species as well as many other genes with another chloroplast *atpB* gene coding H+ -ATPase subunits [92-95]. The *atpB-rbcL* sequence reaches different lengths in *Glycine* as well as in other seed plants. The study by Chiang (1998) shows that the size of the *atpB-rbcL* space in the studied species ranges from 524 bp to 1000 bp [5], where in the non-coding region the occurrence of deletions and insertions, as well as a number of nucleotide substitutions is a common phenomenon, which can also be observed in *Glycine*. In *Glycine max*, its chloroplast genome differs from the core set chloroplast DNA genes because of the presence of a single, large inversion of approximately 51 kb, in the area between the *rbcL* gene and the *rps16* intron [96]. This inversion is also present in other legumes: the mutation was reported in *Lotus* and *Medicago* [96]. In addition, the non-coding *atpB-rbcL* region is rich in AT, due to which most non-coding regions rich in these base pairs show a small number of functions [97, 98]. Therefore, this predisposes them for faster evolution, and hence for use in molecular systematics.

The summary phylogeny was based on sequence of several cpDNA genes from hundreds of spermatophytes including *Glycine* (Table 2). These genes can be divided into three classes. The genes encoding the photosynthetic apparatus structure form the first class. The second class includes the rRNA genes and genes encoding the chloroplast genetic apparatus. The last class consists of an average of about 30 tRNA encoding genes [99], although their number can vary from 20 to 40 [100, 101].

Genes	Products
Genes for the photosynthesis system	
rbcL	Ribulose -1,5- bisphosphate carboxylase, large subunit
psaA, B	Photosystem I, P700 apoproteins A1, A2
psaC	9kDa protein
psaA	Photosystem II, D1 protein
psaB	47kDa chlorophyll a-binding protein
psaC	43 kDa chlorophyll a-binding protein
psaD	D2 protein
psaE	Cytochrome b559 (8kDa protein)
psaF	Cytochrome b559 (4kDa protein)
psaH	10 kDa phosphoprotein

Genes	Products
Genes for the photosynthesis system	
psaI, J, K, L, M, N	–J, -K, -L, -M, -N-proteins
atpA, B, E	H^+-ATPase, CF_1 subunits α, β, ε
atp F, H, I	CF_0 subunits I, III, IV
petB, D	Cytochrome b_6/f complex, subunit b_6, IV
nadA- K	NADH Dehydrogenase, subunits ND 1, NDI 1
Genes for the genetic system	
16S rRNA	16S rRNA
23S rRNA	23S rRNA
trnA -UGC	Alanine tRNA (UGC)
trnG- UCC	Gliycine tRNA (UCC)
rnH- GUG	Histidine tRNA (GUG)
trnI- GAU	Isoleucine tRNA (GAU)
trnK- UUU	Lysine tRNA (UUU)
trnL- UAA	Leucine tRNA (UAA)
rps2, 7, 12, 16	30S: ribosomal proteins CS2, CS7, CS12, CS16
rp12, 20, 32	50S: ribosomal proteins CL2, CL 20, CL32
rpoA, B, C1, C2	RNA polymerase, subunits α, β, β′, β′′
matK	Maturase –like protein
sprA	Small plastid RNA
Others	
clpP	ATP-dependent protease, proteolytic subunit
irf168 (ycf3)	Intron- containing Reading frame (168 codons)

Table 2. Chloroplast genes for the photosynthesis system, for the genetic and others.

The complete size of the *Glycine max* chloroplast genome is 152,218 bp. It contains 25,574 bp of inverted repeats (IRa and IRb), which are separated by a unique small single copy (SSC) region (17,895 bp) [98]. In addition, this genome consists of a large single region (LSC) of unique sequences with 83,175 bp. The IR extends from the *rps19* gene up to the *ycf1*. The *Glycine* chloroplast genome contains 111 unique genes and 19 duplicate copies in the IR, amounting to a total of 130 genes. The cpDNA analysis has showed the presence of 30 different tRNAs in it and 7 of them are repeated within the IR regions. The genes are composed in 60% of encoding regions (52% are protein coding genes and 8% are RNA genes), and in 40% of non-coding regions, including both intergenic spacers and introns. The total content of GC and AT pairs in the *Glycine* chloroplast genes is 34% and 66% respectively. Distinctly higher percentage of AT pairs (70%) was observed in non-coding regions than in coding regions (62% AT) [98].

In comparison with other eukaryotic genomes, cpDNA is highly concentrated, for example, only 32% of the rice genome is non-coding. In *Glycine max* it is slightly more – 40%. Most of the non-coding DNA is found in very short fragments that separate functional genes. Some studies have shown complex patterns of mutational changes in the non-coding regions. Some of the best known regions in the chloroplast genome is the farther region of the *rbcL* gene in many legumes. This non-coding sequence is flanked by the *rbcL* and *psaI* (the gene encoding the polypeptide I of photosystem I).

3.2. Extent of IR in Glycine

Analysis of the IR (inverted repeats) regions in *Glycine max* has shown that they are separated by a large region and a small region of a unique sequence. In cpDNA repeated sequences are usually located asymmetrically, which results in the formation of long and short regions of a unique sequence [102]. The IR in *Glycine* is a region with 25,574 bp containing 19 genes. At the IR/LSC junction, at the ends of the 5' IR, there is the repeated *rps19* gene (68 bp), and at the junction of the IR/SSC and 5' ends the duplicated *ycf1* gene (478 bp) is located. In the course of study it was shown that comparing cpDNA IR region in *Medicago, Lotus, Glycine* and *Arabidopsis* indicates that there are changes within the IR in the two legumes. *Glycine* and *Lotus* have 478 bp and 514 bp of the *ycf1* duplicated, whereas *Arabidopsis* has 1,027 bp duplicated in the IR. This contraction of the IR in these legumes accounts for the smaller size of their IR and larger size of the SSC. In addition, contraction of the IR boundary in legumes, IRa has been lost in *Medicago*. This loss has resulted in *ndhF* (usually located in the SSC) being adjacent to *trnH* (usually the first gene in the LSC at the LSC/IRa junction). Loss of one copy of the IR in some legumes provides support for monophyly of six tribes [103-106]. Wolfe (1988) identified duplicated sequences of portions of two genes, 40 bp of *psbA* and 64 bp of *rbcL*, in the region of the IR deletion between *trnH* and *ndhF* in *Pisum sativum* and these duplications were later identified in broad bean (*Vicia faba*) [104,107]. According to many researchers, the IR region is considered the most conserved part of the chloroplast genome, and thus, it is responsible for stabilizing the plastid DNA molecules [108, 109]. Thus the loss of IR can be phylogenetically informative at the local level, as well as misleading at the global phylogeny level, because the IR loss likely occurred independently in more than one group of plants. Coniferous and some legumes (*Pisum sativum, Vicia faba, Medicago sativa*), for example, contain only one IR. Perhaps the lack of repeat sequences in these plants is associated with an increased incidence of rearrangement of chloroplast genomes [109].

Introns or intergenic sequences in legume chloroplast DNA have become extremely important tools in phylogenetic analyses aimed at systematizing of this species [110, 111]. Moreover, their microstructural changes occur with great frequency in the regions of cpDNA. The body of existing research suggests that mutations in the non-coding regions and relatively fast evolution of the organelle genome encoding regions can serve as valuable markers for the separation species in their evolutionary origin [110, 111]. The systematics of plants generally considers chloroplast indeles to be phylogenetic markers, because of their low prevalence in comparison with nucleotide substitutions [5].

3.3. CpDNA markers

There are many methods of generating molecular markers that rely on site-specific amplification of a selected DNA fragment using polymerase chain reaction (PCR) and its further processing (restriction analysis, sequencing). Initially the research on the plant genome (mostly phylogenetic studies) used non-coding and coding sequences of chloroplast DNA. With time, the genes or DNA segments located in the nuclear DNA, mitochondrial (mtDNA) and chloroplast (cpDNA) found a prominent place among plant DNA markers. Fully automated DNA sequencing made it possible to subject ever-newer regions of plant DNA to comparative sequencing.

One of the most frequently sequenced cpDNA fragments in plant phylogeny of spermatophytes is the *rbcL* gene encoding a large ribulose bisphosphate carboxylase subunit (RUBISCO), whose length in most plants is 1,428, 1,431 and 1,434 bp, and insertions and deletions within it are extremely rare [94]. For many years this gene has been the subject of many comprehensive phylogenetic analyses of subgenus *Glycine* [112-114]. The *rbcL* is most commonly used in the analyses at the family and genus levels, but there also exists research at the lower levels, cultivars and wild soybean [98, 115, 116]. A marker with very similar characteristics to those of the *rbcL* (the rate of evolution, the length of 1497 bp) is a gene encoding the ATP synthase β subunit – the *atpB* [94].The *matK* gene sequence, encoding maturase involved in splicing of the type II introns, and whose length is 1,550 bp is characterized by a rapid rate of evolution that allows to use it in research at the species and genus levels [117, 118]. Frequent mutations in this gene make it unsuitable for studies at higher taxonomic levels. Other popular cpDNA sequences used in phylogenetic studies of legumes include the *ndhF* (the gene encoding the NADH protein, which is a dehyd98rogenase subunit), 16S rDNA, the non-coding *atpB-rbcL* region [94], or the *trnL* (UAA) intron and mediator between the *trnL* (UAA) exon and the *trnF* (GAA) gene [96, 117- 119].

It should be noted that the rate of evolution for a specific DNA region to be used as a marker can vary significantly not only among systematic groups, but also within these groups [98]. Moreover, each DNA fragment within the same group has a different rate of evolution, such as the *ndhF* cpDNA sequence in the *Solanaceae* family, which provides about 1.5 times more information in terms of parsimony than the *rbcL* [90]. Therefore each gene or any other DNA fragment used as a genetic marker has a typical range of "taxonomic" or phylogenetic applications, which can vary significantly within a taxon. For this reason, the *rbcL* sequence has been widely used in *Gycine* for many years at the species and genus levels [104, 117, 118].

3.4. The genetic diversity of soybeans

The importance of genetic variations in facilitating plant breeding and/or conservation strategies has long been recognized [121]. Molecular markers are useful tools for assaying genetic variation and provide an efficient means to link phenotypic and genotypic variation [122].

In recent years, the progress made in the development of DNA based marker systems has advanced our understanding of genetic resources. These molecular markers are classified as: (i) hybridization based markers i.e. restriction fragment length polymorphisms (RFLPs), (ii) PCR-based markers i.e. random amplification of polymorphic DNAs (RAPDs), amplified fragmentlength polymorphisms (AFLPs), inter simple sequence repeats (ISSRs) and microsatellites or simple sequence repeats (SSRs), and (iii) sequence based markers i.e. single nucleotide polymorphisms (SNPs) [121, 123]. Majority of these molecular markers have been developed either from genomic DNA library (e.g. RFLPs or SSRs) or from random PCR amplification of genomic DNA (e.g. RAPDs) or both (e.g. AFLPs) [123]. Availability of an array of molecular marker techniques and their modifications led to comparative studies among them in many crops including soybean, wheat and barley [124-126]. Among all these, SSR markers have gained considerable importance in plant genetics and breeding owing to many desirable attributes including hypervariability, multiallelic nature, codominant inheritance, reproducibility, relative abundance, extensive genome coverage (including organellar genomes), chromosome specific location, amenability to automation and high throughput genotyping [127]. In contrast, RAPD assays are not sufficiently reproducible whereas RFLPs are not readily adaptable to high throughput sampling. AFLP is complicated as individual bands are often composed of multiple fragments mainly in large genome templates [123]. The general features of DNA markers are presented in Table 3.

	Molecular markers			
	EST–SSRs	SSRs	RFLPs	RAPDs/AFLPs/ISSRs
Need for sequence data	Essential	Essential	Not required	Not required
Level of polymorphism	Low	High	Low	Low-moderate
Dominance	Co-dominant	Co-dominant	Co-dominant	Co-dominant
Interspecific transferability	High	Low-moderate	Moderate-high	Low-moderate
Utility in Marker assisted selection	High	High	Moderate	Low-moderate
Cost and labour involved in generation	Low	High	High	Low-moderate

Table 3. Important features of different types of molecular markers.

The genetic diversity of wild and cultivated soybeans has been studied by various techniques including isozymes [128], RFLP [87], SSR markers [124], and cytoplasmic DNA markers [87, 128, 129]. Based on haplotype analysis of chloroplast DNA, cultivated soybean appears to have multiple origins from different wild soybean populations [129, 130].

Using PCR-RFLP method soybean chloroplast DNAs were classified into three main haplotype groups (I, II and III) [113, 130, 131]. Type I is mainly found in the species of cultivated soybean (*Glycine max*), while types II and III are often found in both the cultivated and wild forms of soybean (*Glycine soja*). Type III is by far the most dominant in the wild soybean species [113]. In *Glycine*, these types are widely used in evaluating cpDNA variability and in determining phylogenetic relationships between different types of cpDNA using different marker systems. According to Chen and Hebert (1999) [133] analysis of cpDNA sequence is not sufficient for when the analysis of population genetics, and so cpDNA polymorphism assessment methods must be constantly complemented with methods such as single-strand conformation polymorphism (SSCP) [134], or dideoxy fingerprinting (ddF) [135], and directed termination and polymerase chain reaction (DT-PCR). However, some researchers point out that there are many disadvantages of these methods, mainly because of their high cost and large amount of work necessary for obtaining the results. In their view, a single change in the regions of *Glycine* chloroplast DNA at the species and genus levels should be located on a local-specific markers, for example, non-coding regions, using PCR and sequencing.

Analyses of non-coding regions of cpDNA have been employed to elucidate phylogenetic relationship of different taxa [90]. Compared with coding regions, non-coding regions may provide more informative characters in phylogenetic studies at the species level because of their high variability due to the lack of functional constraints. Non-coding regions of cpDNA have been assayed either by direct sequencing [136-141], or by restriction-site analysis of PCR products (PCR-RFLP) [142-146]. In Small's opinion (1998) non-coding regions, which include introns and intergenic sequences, often show greater variability at nucleotides than at the encoding regions, which makes the non-coding regions good phylogenetic markers [139]. Mutations in the form of insertions and deletions are accumulated in noncoding regions at the same rate as nucleotide substitutions, and such kinds of mutations significantly accelerate changes in these regions. In many cases, insertions or deletions are related to short repeat sequences. Therefore, many researchers continually focus on the analysis of non-coding regions. Using RFLP method, Close et al. (1989) found six cpDNA haplotypes and described them in types, ranging from group I to VI, including cultivated and wild soybeans [147]. In the course of their research they found that groups I and II diverge from groups III to VI, thus dividing subgenus *Soja* into two main groups. They presented a hypothesis that group II can be distinguished form group III by two independent mutations. Similar groups of haplotypes in legumes were also obtained by Shimamoto et al, (1992) [128] and Kanazawa et al, (1998) [148], using a combination of *Eco*RI and *Cla*I RFLPs. In their classification, Kazanawa et al. (1998) relied on sequential analysis and found that differences in the three types described by Shimamoto et al. (1992) resulted from two single-base substitutions: one in the non-coding region, between the *rps11 and rpl36,* and the other in the 3' part of the coding region of the *rps3*. Based on the existing reports, Xu et al. (2000) sequenced nine non-coding regions of cpDNA for seven cultivars and 12 wild forms of soybean (*Glycine max, Glycine soja, Glycine tabacina, Glycine tomentella, Glycine microphylla, Glycine clandestina*) in order to verify earlier classification of *Glycine* [113]. In the course of their studies, they located eleven single-base changes (substitutions and deletions) in the collected 3849 database. They located five mutations in the distinguished haplotypes I and II, and seven mutations in type III. In addition, haplotypes I and II were identical and clearly different from the taxons in type III. This research has not yielded significant results, because different types of cpDNA could not originate monophyletically, but it contributed to finding a common ancestor in the course of evolution of *Glycine*. A neighbor joining tree resulting from the sequence data revealed that the subgenus *Soja* connected with *Glycine microphylla,* which formed a distinct clad from *Clycine clandestine* and the tetraploid cytotypes of *Glycine tabacina* and *Glycine tomentella.* Several informative length mutations of 54 to 202 bases, due to insertions or deletions, were also detected among the species of the genus *Glycine.*

3.5. Non-coding regions of the chloroplast genome as site-specific markers in Glycine

In the chloroplast genomes of legumes, including soybean, there are many non-coding regions, which are characterized by a faster rate of evolution when compared to the coding regions. As mentioned earlier some of the chloroplast genes have introns, yet their structure differs from those occuring in the nuclear genes, since in the case of cpDNA introns have a tendency to adopt secondary structure, which affects the model in which cpDNA introns evolve and it is enforeced by the secondary structure. This restriction in changes caused by mutations affects the functional requirements related to the formation of introns [98, 108]. As there are no adequate studies on the evolution of introns, it can be assumed that their evolution is similar to that of the protein-encoding genes. The loss of introns in the course of the evolution of chloroplast DNA is an interesting process. It has been discovered that *O. sativa* has 3 introns less in cpDNA than *M. polymorpha* and *N. tabacum.* The loss of an intron in the *rpl2* gene was researched in 340 species representing 109 families of angiosperms including *Glycine* [149]. When trying to determine the taxonomic position, the absence of this intron in a given gene shows that it was lost at least six times in the evolution of angiosperms. In *Glycine* 23 introns have been identified while in *Arabidopsis thaliana* there are 26 introns, mostly located in the same genes and in the same locations within those genes [98, 102].

Non-coding regions in chloroplast DNA have become a major source for phylogenetic studies within the species *Glycine* and in many other seed plants. Earlier, the most popular phylogeny sequences included encoding regions, such as the *rbcL* gene sequences that were designed to determine the phylogenetic relationships between species in major taxonomic groups [113, 136-141]. According to Taberlet et al. (1991) [119] the potential ability of noncoding regions of cpDNA was reserved for species located in the lower taxonomic levels while the non-coding regions, which include introns and intergenic sequences, often show greater variability at nucleotides than is evident in the coding regions, which predisposes them to be used in population studies involving *Glycine,* and others [139,142].

As the result, many studies on phylogenetic utility of non-coding regions have been published [110]. For example [150]: *trnH-psbA, trnS-tang;* [148]: *rps11-rpl36, rpl16-rps3,* [113]: *trnT-trnL.*

In cpDNA analysis of many plants, very conservative regions flanking areas with high variability are used. The more conservative regions, the higher the chance for the primers designed in the PCR reaction, which will be able to join the broader taxonomic group [96, 113]. The region occurring between the *trnT* (UGU) and the *trnF* (GAA) genes is a large single copy wich is suitable because of the conservativeness of the *trn* genes and several hundred base pairs of noncoding regions. The intergenic space between the *trnT* (UGU) and the *trnL* (UAA) 5' exon ranged from 298 bp to about 700 bp in the species studied by [119]. In the plant genomes completely sequenced by Sugiute, the length of this region is different and amounts to 770 bp in rice and 710 bp in tobacco. In *Marchantia polymorpha* it is 188 bp [151]. This region is located between the tRNA genes, just as the non-coding sequence located between the *trnL* (UAA) 3' exon and the *trnF* (GAA). Due to its catalytic properties and its secondary structure, the *trnL* (UAA) intron, which belongs to type I introns, is less variable and therefore of better utility for evolutionary studies at higher taxonomic levels [113]. Moreover, depending on the species, they show high frequency of insertions or deletions, which makes them potentially useful as genetic markers.

Region	Primer sequence (5 - 3)*	Annealing temperature	Reference
trnH–psbA	f:TGATCCACTTGGCTACATCCGCC	60°C	[99] (tobacco)
	r: GCTAACCTTGGTATGGAAGT		[150] (soybean)
trnS–trnG	f: GATTAGCAATCCGCCGCTTT	60°C	[99] (tobacco)
	r: TTACCACTAAACTATACCCGC		[99] (tobacco)
trnT–trnL	f: GGATTCGAACCGATGACCAT	60°C	[113] (soybean)
	r: TTAAGTCCGTAGCGTCTACC		[99] (tobacco)
trnL–trnF	f: TCGTGAGGGTTCAAGTCC	56°C	[99] (tobacco)
	r: AGATTTGAACTGGTGACACG		[99] (tobacco)
atpB–rbcL	f: GAAGTAGTAGGATTGATTCTC	58°C	[99] (tobacco)
	r: CAACACTTGCTTTAGTCTCTG		[99] (tobacco)
psbB–psbH	f: AGATGTTTTTGCTGGTATTGA	56°C	[99] (tobacco)
	r: TTCAACAGTTTGTGTAGCCA		[99] (tobacco)
rps11–rpl36	f:GTATGGATATATCCATTTCGTG	50°C	[148] (soybean)
	r: TGAATAACTTACCCATGAATC		[148] (soybean)
rpl16–rps3	f: ACTGAACAGGCGGGTACA	50°C	[148] (soybean)
	r: ATCCGAAGCGATGCGTTG		[148] (soybean)
ndhD–ndhE	f: GAAAATTAAGGAACCCGCAA	56°C	[99] (tobacco)
	r: TCAACTCGTATCAACCAATC		[99] (tobacco)

*f, forward primer; r, reverse primer

Table 4. Primers used for amplification of nine non-coding regions of soybean cpDNA.

In most studied species, the *trnL* (UAA) intron ranges in size from 254 - 767 bp. Its smaller fragment – the P6 loop – reaches a length of 10 - 143 bp. It is commonly applied in DNA barcoding. Its main limitation lies in its low homologousness with the species from the Gene Bank, which amounts to 67.3%, while the homologousness of the P6 loop is 19.5%. However, it also has some advantages: conservative primers projected form and trouble-free amplification process. Amplification of the P6 loop can be performed even in a very degraded DNA. The intron is well known and its sequences are used to determine phylogenetic relationships between closely related species or to identify a plant species [152]. The first universal primers for this region were designed more than 20 years ago [119]. However, it does not belong to the most variable non-coding regions in chloroplast DNA [108]. The *trnL* (UAA) intron is the only one belonging to group I introns in chloroplast DNA, which means that its secondary structure is highly conservative, with a possibility of changes in its conservative [113] and variability in regions [99, 153]. Consequently, comparing the diversity of the *trnL* intron sequences allows to obtain new primers that contain conservative regions and amplify short sections contained between them [152].

Thus, in angiosperms, using non-coding regions in research at lower levels of the genome is a routine practice [108]. A large number of non-coding regions of cpDNA has been located in angiosperms, some of which are highly variable, whereas others show relatively small variability [108]. In studying the chloroplast genome, many researchers looked for universal primers that would allow amplification of many non-coding regions of cpDNA (Table 4) [111, 113, 148, 150].

4. Conclusion

In phylogenetic and population studies of *Glycine*, genetic information contained not only in cpDNA but also in mtDNA are often analysed. Organelle DNA can be used to find species-specific molecular markers. Molecular markers are an important tool to systematize the species because their use allows for detecting the differences in the genes directly. The selection of appropriate sequences, which depends on the taxonomic level at which reconstruction of the origin is carried out is very important. The initial selection concerns non-conservative sequences, which are subject to fast evolution, because the more related the specimen are, the more changeable the region should be. The relatively slow rate of evolution of certain sequences may exclude statistically significant analyses within families or species, while the study of relationship between species, which phylogenetically are very distant, using more slowly evolving sequences can be very useful. Non-coding sequences show a faster rate of evolution than the coding sequences. These regions accumulate a greater number of insertion/deletion or substitution than the non-coding regions, and therefore may be more suitable for research at inter-or intra-genus levels.

Author details

Lidia Skuza*, Ewa Filip and Izabela Szućko

*Address all correspondence to: skuza@univ.szczecin.pl

Cell Biology Department, Faculty of Biology, University of Szczecin, Poland

References

[1] Petit, R. J., Duminil, J., Fineschi, S., Hampe, A., Salvini, D., & Vendramin, G. G. (2005). Comparative organization of chloroplast, mitochondrial and nuclear diversity in plant populations. *Molecular Ecology*, 14, 689-701.

[2] Wagner, D. B. (1992). Nuclear, chloroplast, and mitochondrial DNA polymorphisms as biochemical markers in population genetic analyses of forest trees. *New Forests*, 6, 373-390.

[3] Liepelt, S., Bialozyt, R., & Ziegenhagen, B. (2002). Wind-dispersed pollen mediates postglacial gene flow among refugia. USA. *Proceedings of the Natlional Academy of Sciences*, 99, 14590-14594.

[4] Johansen, A. D., & Latta, R. G. (2003). Mitochondrial haplotype distribution, seed dispersal and patterns of postglacial expansion of ponderosa pine. *Molecular Ecology*, 12, 293-298.

[5] Chiang, Y. C., Hung, K. H., Schaal, B. A., Ge, X. J., Hsu, T. W., & Chiang, T. Y. (2006). Contrasting phylogeographical patterns between mainland and island taxa of the Pinus luchuensis complex. *Molecular Ecology*, 15, 765-779.

[6] Jaramillo-Correa, J. P., Beaulieu, J., & Bousquet, J. (2004). Variation in mitochondrial DNA reveals multiple distant glacial refugia in black spruce (Picea mariana), a transcontinental North American conifer. *Molecular Ecology*, 13, 2735-2747.

[7] Godbout, J., Jaramillo, J. P., Beaulieu, J., & Bousquet, J. (2005). A mitochondrial DNA minisatellite reveals the postglacial history of jack pine (Pinus banksiana), a broad-range North American conifer. *Molecual Ecology*, 14, 3497-3512.

[8] Huang, S., Chiang, Y. C., Schaal, B. A., Chou, C. H., & Chiang, T. Y. (2001). Organelle DNA phylogeography of Cycas taitungensis, a relict species in Taiwan. *Molecular Ecology*, 10, 2669-2681.

[9] Pleines, T., Jakob, S. S., & Blattner, D. B. (2009). Application of non-coding DNA regions in intraspecific analyses. *Plant Systematic and Evolution*, 282, 281-294.

[10] Tautz, D. (1989). Hypervariability of simple sequences as a general source for polymorphic DNA markers. *Nucleic Acids Research*, 17, 6463-6471.

[11] Tautz, D., Trick, M., & Dover, G. A. (1986). Cryptic simplicity in DNA is a major source of genetic variation. *Nature*, 322, 652-656.

[12] Lagercrantz, U., Ellegren, H., & Andersson, L. (1993). The abundance of various polymorphic microsatellite motifs differs between plants and vertebrates. *Nucleic Acids Research*, 21, 1111-1115.

[13] Powell, W., Morgante, M., Mc Devitt, R., Vendramin, G. G., & Rafalski, J. A. (1995). Polymorphic simple sequence repeat regions in chloroplast genomes: applications to

the population genetics of pines. USA. *Proceedings of the Natlional Academy of Sciences*, 92, 7759-77763.

[14] Birky, C. W., Fuerst, P., & Maruyama, T. (1989). Organelle gene diversity under migration, mutation, and drift: equilibrium expectations, approach to equilibrium, effects of heteroplasmic cells, and comparison to nuclear genes. *Genetics*, 121, 613-627.

[15] Rendell, S., & Ennos, R. A. (2002). Chloroplast DNA diversity in Calluna vulgaris (heather) populations in Europe. *Molecular Ecology*, 11, 69-78.

[16] Hudson, R. R., & Coyne, J. A. (2002). Mathematical consequences of the genealogical species concept. *Evolution*, 56, 1557-1565.

[17] Kadereit, J. W., Arafeh, R., Somogyi, G., & Westberg, E. (2005). Terrestrial growth and marine dispersal? Comparative phylogeography of five coastal plant species at a European scale. *Taxon*, 54, 861-876.

[18] Avise, J. C. (2000). *Phylogeography: the history and formation of species*, Cambridge, Harvard University Press.

[19] Richards, M. B., Macauly, V. A., Bandelt, H. J., & Sykes, B. C. (1998). Phylogeography of mitochondrial DNA in western Europe. *Annals of Human Genetics*, 62, 241-260.

[20] Tomaru, N., Takahashi, M., Tsumura, Y., Takahashi, M., & Ohba, K. (1998). Intraspecific variation and phylogeographic patterns of Fagus crenata (Fagaceae) mitochondrial DNA. *American Journal of Botany*, 85, 629-636.

[21] Mackenzie, S., & Mc Intosh, L. (1999). Higher plant mitochondria. *Plant Cell*, 11, 571-585.

[22] Schuster, W., & Brennicke, A. (1994). The plant mitochondrial genome: physical structure, information content, RNA editing, and gene migration to the nucleus. *Annual Review of Plant Physiology and Plant Molecular Biology*, 45, 61-78.

[23] Palmer, J. D. (1992). Mitochondrial DNA in plant systematics: applicationsand limitations. *In: Soltis PS, Soltis DE, Doyle JJ. (ed.) Molecular systematics of plants*, New York, Chapman and Hall, 26-49.

[24] Wolfe, K. H., Li, W. H., & Sharp, P. M. (1987). Rates of nucleotide substitution vary greatly among plant mitochondrial, chloroplast, and nuclear DNAs. USA. *Proceedings of the Natlional Academy of Sciences*, 84, 9054-9058.

[25] Larsson, N. G., & Clayton, D. A. (1995). Molecular genetic aspects of human mitochondrial disease. *Annual Review of Genetics*, 29, 151-178.

[26] Binder, S., Marchfelder, A., & Brennicke, A. (1996). Regulation of gene expression in plant mitochondria. *Plant Molecular Biology*, 32, 303-314.

[27] Unseld, M., Marienfeld, J., Brandt, P., & Brennicke, A. (1997). The mitochondrial genome of Arabidopsis thaliana contains 57 genes in 366,924 nucleotides. *Nature Genetics*, 15, 57-61.

[28] Oda, K., Yamato, K., Ohata, E., Nakamura, Y., Takemura, M., Nozato, N., Akashi, K., Kanrgae, T., Ogura, Y., Kohchi, T., & Ohyama, K. (1992). Gene organization from the complete sequence of liverwort, Marchantia polymorpha, mitochondrial DNA: a primitive form of plant mitochondrial genome. *Journal of Molecular Biology*, 223, 1-7.

[29] Backert, S., Nielsen, B. L., & Borner, T. (1997). The mystery of the rings: structure and replication of mitochondrial genomes from higher plants. *Trends in Plant Science*, 2, 477-483.

[30] Andre, C. P., & Walbot, V. (1995). Pulsed-field gel mapping of maize mitochondrial chromosomes. *Molecular and General Genetics*, 247, 255-263.

[31] Bendich, A. J. (1993). Reaching for the ring: the study of mitochondrial genome structure. *Current Genetetics*, 24, 279-290.

[32] Fauron, C., Casper, M., Gao, Y., & Moore, B. (1995). The maize mitochondrial genome: dynamic, yet functional. *Trends in Genetics*, 11, 228-235.

[33] Fauron, C. M. R., Casper, M., Gesteland, R., & Albertsen, M. (1992). A multi-recombination model for the mtDNA rearrangements seen in maize cmsT regenerated plants. *The Plant Journal*, 2, 949-958.

[34] Palmer, J. D., & Herbon, L. A. (1988). Plant mitochondrial DNA evolves rapidly in structure, but slowly in sequence. *Journal of Molecular Evolution*, 28, 87-97.

[35] Palmer, J. D., & Shields, C. R. (1984). Tripartite structure of the Brassica campestris mitochondrial genome. *Nature*, 307, 437-440.

[36] Palmer, J. D., Makaroff, C. A., Apel, I. J., & Shirzadegan, M. (1990). Fluid structure of plant mitochondrial genomes: evolutionary and functional implications. *In: Clegg MT, O'Brien SJ (ed.) Molecular Evolution*, New York, Alan R. Liss, 85-96.

[37] Mosig, G. (1987). The essential role of recombination in phage T4 growth. *Annual Review of Genetics*, 21, 347-371.

[38] Mosig, G. (1998). Recombination and recombinationdependent DNA replication in bacteriophage T4. *Annual Review of Genetics*, 32, 379-413.

[39] Oldenburg, D. J., & Bendich, A. J. (2001). Mitochondrial DNA from the liverwort Marchantia polymorpha: circularly permuted linear molecules, head-to-tail concatemers, and a 50 protein. *Journal of Molecular Biology*, 310, 549-562.

[40] Backert, S., Dorfel, P., & Borner, T. (1995). Investigation of plant organellar DNA by pulsed-field gel electrophoresis. *Current Genetics*, 28, 390-399.

[41] Backert, S., Lurz, R., Oyarzabal, O. A., & Borner, T. (1997). High content, size and distribution of singlestranded DNA in the mitochondria of Chenopodium album (L.). *Plant Molecular Biology*, 33, 1037-1050.

[42] Bendich, A. J., & Smith, S. B. (1990). Moving pictures and pulsed-field gel electrophoresis show linear DNA molecules from chloroplasts and mitochondria. *Current Genetics*, 17, 421-425.

[43] Bendich, A.J. (1996). Structural analysis of mitochondrial DNA molecules from fungi and plants using moving pictures and pulsed-field gel electrophoresis. *Journal of Molecular Biology*, 225, 564-588.

[44] Oldenburg, D. J., & Bendich, A. J. (1996). Size and structure of replicating mitochondrial DNA in cultured tobacco cells. *Plant Cell*, 8, 447-461.

[45] Sena, E. P., Revet, B., & Moustacchi, E. (1986). In vivo homologous recombination intermediates of yeast mitochondrial DNA analyzed by electron microscopy. *Molecular and General Genetics*, 202, 421-428.

[46] Palmer, J. D., & Herbon, L. A. (1987). Unicircular structure of the Brassica hirta mitochondrial genome. *Current Genetics*, 11, 565-570.

[47] Stern, D. B., & Palmer, J. D. (1984). Recombination sequences in plant mitochondrial genomes: Diversity and homologies to known mitochondrial genes. *Nucleic Acid Research*, 12, 6141-6157.

[48] Chanut, F. A., Grabau, E. A., & Gesteland, R. F. (1993). Complex organization of the soybean mitochondrial genome: recombination repeats and multiple transcripts at the atpA loci. *Current Genetics*, 23, 234-247.

[49] Dewey, R. E., Levings, C. S., III., & Timothy, D. H. (1986). Novel recom- binations in the maize mitochondrial genome produce a unique transcriptional unit in the Texas male-sterile cytoplasm. *Cell*, 44, 439.

[50] Laver, H. K., Reynolds, S. J., Moneger, F., & Leaver, C. J. (1991). Mitochondrial genome organization and expression associated with cytoplasmic male sterility in sunflower (Helianthus annuus). *Plant Journal*, 1, 185-193.

[51] Johns, C., Lu, M., Lyznik, A., & Mackenzie, S. (1994). A mitochondrial DNA sequence is associated with abnormal pollen development in cytoplasmic male sterile bean plants. *Plant Cell*, 4, 435-449.

[52] Martinez-Zapater, J. M., Gil, P., Capel, J., & Somerville, C. R. (1992). Mutations at the Arabidopsis CHM locus promote rearrangements of the mitochondrial genome. *Plant Cell*, 4, 889-899.

[53] Newton, K., Knudsen, C., Gabay-Laughnan, S., & Laughnan, J. (1990). An abnormal growth mutant in maize has a defective mitochondrial cytochrome oxidase gene. *Plant Cell*, 2, 107-113.

[54] Levings, C. S., III., & Pring, D. R. (1979). Mitochondrial DNA of higher plants and genetic engineering. In: Setlow JK, Hollaender A (ed.). *Genetic engineering*, 1, New York, Plenum Press, 205-222.

[55] Bailey-Serres, J., Leroy, P., Jones, S. S., Wahleithner, J. A., & Wolstenholme, D. R. (1987). Size distribution of circular molecules in plant mitochondrial DNAs. *Current Genetics*, 12, 49-53.

[56] Grabau, E., Davis, W. H., & Gengenbach, B. G. (1989). Restriction fragment length polymorphism in a subclass of the mandarin soybean. *Crop Science*, 29, 1554-1559.

[57] Synenki, R. M., Levings, C. S., III., & Shah, D. M. (1978). Physicochemical characterization of mitochondrial DNA from soybean. *Plant Physiology*, 61, 460-464.

[58] Kanazawa, A., Tozuka, A., Kato, S., Mikami, T., Abe, J., & Shimamoto, Y. (1998). Small interspersed sequences that serve as recombination sites at the cox2 and atp6 loci in the mitochondrial genome of soybean are widely distributed in higher plants. *Current Genetics*, 33, 188-198.

[59] Kato, S., Kanazawa, A., Mikami, T., & Shimamoto, Y. (1998). Evolutionary changes in the structures of the cox2 and atp6 loci in the mitochondrial genome of soybean involving recombination across small interspersed sequences. *Current Genetics*, 34, 303-312.

[60] Grabau, E. A., Davis, W. H., Phelps, N. D., & Gengenbach, B. G. (1992). Classification of soybean cultivars based on mitochondrial DNA restriction fragment length polymorphisms. *Crop Science*, 32, 271-274.

[61] Moeykens, C. A., Mackenzie, S. A., & Shoemaker, R. C. (1995). Mitochondrial genome diversity in soybean: repeats and rearrangements. *Plant Molecular Biology*, 29, 245-254.

[62] Hanlon, R. W., & Grabau, E. A. (1997). Comparison of mitochondrial organization of four soybean cytoplasmic types by restriction mapping. *Soybean Genetics Newsletter*, 24, 208-210.

[63] Khazi, F. R., Edmondson, A. C., & Nielsen, B. L. (2003). An Arabidopsis homologue of bacterial RecA that complements an E. coli recA deletion is targeted to plant mitochondria. *Molecular and General Genetics*, 269, 454-463.

[64] Manchekar, M., Scissum-Gunn, K., Song, D., Khazi, F., Mc Lean, S. L., & Nielsen, B. L. (2006). DNA Recombination Activity in Soybean Mitochondria. *Journal of Molecular Bioliology*, 356, 288-299.

[65] Grabau, E. A., Havlik, M., & Gesteland, R. F. (1988). Chimeric Organization of Two Genes for the Soybean Mitochondrial Atpase Subunit 6. *Current Genetics*, 13, 83-89.

[66] Wissinger, B., Brennicke, A., & Schuster, W. (1992). Regenerating good sense: RNA editing and trans-splicing in plant mitochondria. *Trends in Genetics*, 8, 322-328.

[67] Hanlon, R., & Grabau, E. A. (1995). Cytoplasmic diversity in old domestic varieties of soybean using two mitochondrial markers. *Crop Science*, 35, 1148-1151.

[68] Grabau, E. A., & Davis, W. H. (1992). Cytoplasmic diversity in Glycine soja. *Soybean Genetics Newslett*, 19, 140-144.

[69] Tozuka, A., Fukushi, H., Hirata, T., Ohara, M., Kanazawa, A., Mikami, T., Abe, J., & Shimamoto, Y. (1998). Composite and clinal distribution of Glycine soja in Japan revealed by RFLP analysis of mitochondrial. *DNA Theoretical and Applied Genetics*, 96, 170-176.

[70] Schuster, W., & Brennicke, A. (1987). Nucleotide sequence of the Oenothera ATPase subunit 6 gene. Nucleic Acids Research ., 15, 9092.

[71] Braun, C. J., & Levings, C. S., III. (1985). Nucleotide Sequence of the F(1)-ATPase alpha Subunit Gene from Maize Mitochondria. *Plant Physiology*, 79(2), 571-577.

[72] Isaac, G. I., Brennicke, A., Dunbar, M., & Leaver, C. J. (1985). The mitochondrial genome of fertile maize (Zea mays L.) contains two copies of the gene encoding the alpha-subunit of the F1-ATPase. *Current Genetics*, 10(4), 321-328.

[73] Schuster, W., & Brennicke, A. (1986). Pseudocopies of the ATPase a-subunit gene in Oenothera mitochondria are present on different circular molecules. *Molecular and General Genetics*, 204, 29-35.

[74] Morikami, A., & Nakamura, K. (1987). *Structure and expression of pea mitochondrial F1ATPase alpha-subunit gene and its pseudogene involved in homologous recombination, J Biochem*, 101, 967-976.

[75] Chaumont, F., Boutry, M., Briquet, M., & Vassarotti, A. (1988). Sequence of the gene encoding the mitochondrial F1ATPase alpha subunit from Nicotiana plumbaginifolia. *Nucleic Acids Research*, 16(13), 6247.

[76] Schulte, E., Staubach, S., Laser, B., & Kück, U. (1989). Wheat mitochondrial DNA: organization and sequences of the atpA and atp9 genes. *Nucleic Acids Research*, 17(18), 7531.

[77] Kadowaki, K., Kazama, S., & Suzuki, S. (1990). Nucleotide sequence of the F,-ATPase a subunit genes from rice mitochondria. *Nucleic Acids Research*, 18, 1302.

[78] Köhler, R. H., Lössl, A., & Zetsche, K. (1990). Nucleotide sequence of the F1ATPase alpha subunit gene of sunflower mitochondria. *Nucleic Acids Research*, 18, 4588.

[79] Makaroff, C. A., Apel, I. J., & Palmer, J. D. (1990). Characterization of radish mitochondrial atpA: influence of nuclear background on transcription of atpA-associated sequences and relationship with male sterility. *Plant Molecular Biology*, 15(5), 735-746.

[80] Siculella, L., D'Ambrosio, L., De Tuglie, A. D., & gauerani, R. (1990). Minor differences in the primary structures of atpa genes coded on the mt DNA of fertile and male sterile sunflower lines. *Nucleic Acids Research*, 18(15), 4599.

[81] Handa, H., & Nakajima, K. (1991). Nucleotide sequence and transcription analyses of the rapeseed (Brassica napus L.) mitochondrial F1ATPase alpha-subunit gene. *Plant Molecular Biology*, 16, 361-364.

[82] Chase, C. D., & Ortega, V. M. (1992). Organization of ATPA coding and 3' flanking sequences associated with cytoplasmic male sterility in Phaseolus vulgaris. L. *Current Genetics*, 22, 147-153.

[83] Zurawski, G., Bottomley, W., & Whitfeld, P. R. (1982). Structures of the genes for the beta and c subunits of spinach chloroplast ATPase indicates a dicistronic mRNA and an overlapping translation stop/start signal. USA. *Proceedings of the Natlional Academy of Sciences*, 79, 6260-6264.

[84] Krebbers, E. T., Larrinua, I. M., Mc Intosh, L., & Bogorad, L. (1982). The maize chloroplast genes for the beta and epsilon subunits of the photosynthetic coupling factor CF1 are fused. *Nucleic Acids Research*, 10(16), 4985-5002.

[85] Walker, J. E., Fearnley, I. M., Gay, N. J., Gibson, B. W., Northrop, F. D., Powell, S. J., Runswick, M. J., Saraste, M., & Tybulewicz, V. L. J. (1985). Primary structure and subunit stoichiometry of F1-ATPase from bovine mitochondria. *Journal of Molecular Biology*, 184(4), 677-701.

[86] Shimamoto, Y., Fukushi, H., Abe, J., Kanazawa, A., Gai, J., Gao, Z., & Xu, D. (1998). RFLPs of chloroplast and mitochondrial DNA in wild soybean, Glycine soja, growing in China. *Genetic Resources and Crop Evolution*, 45, 433-439.

[87] Abe, J. (1999, 13th-15th October). The genetic structure of natural populations of wild soybeans revealed by isozymes and RFLP's of mitochondrial DNS's: possible influence of seed dispersal, cross pollination and demography. Japan. *In: The Seventh Ministry of Agriculture, Forestry and Fisheries (MAFF), Japan International Workshop on Genetic Resources Part I Wild Legumes*, National Institute of Agrobiological Resources Tsukuba, Ibaraki, 143-159.

[88] Shimamoto, Y. (2001). Polymorphism and phylogeny of soybean based on chloroplast and mitochondrial DNA analysis. *Japan Agricultural Research Quarterly (JARQ)*, 35(2), 79-84.

[89] Cleeg, M. T., & Zurawski, G. (1992). Chloroplast DNA and the study of plant phylogeny. New York, *In: Soltis PS, Soltis DE, Doyle JJ (eds.) Present status and future procpects in Molecular Systematics of Plants*, Chapman &Hall, 1-13.

[90] Olmstead, R. G., & Palmer, J. D. (1994). Chloroplast DNA systematics: a review of method and data analysis. *American Journal of Botany*, 81, 1205-1224.

[91] Chase, M., & 41others, . (1993). Phylogenetics of seed plants: An analysis of nucleotide sequences from the plastid gene rbcL. *Annals Missouri Botany Gardens*, 80, 528-580.

[92] Parkinson, C. L., Adams, L., & Palmer, D. (1999). Multigene analyses identify the three earliest lineages of extant flowering plants. *Current Biology*, 9, 1485-1488.

[93] Qui, Y. L. (1999). The earliest angiosperms: Evidence from mitochondrial, plastid, and nuclear genomes. *Nature*, 402, 404-407.

[94] Soltis, D. E., Soltis, P. S., & Doyle, J. J. (1998). Molecular systematic of Plants II. *DNA sequencing*, Boston MA, Kluwer.

[95] Soltis, P. S., Soltis, D. E., & Chase, M. W. (1999). Angiosperm phylogeny inferred from multiple genes as a tool for comparative biology. *Nature*, 402, 402-404.

[96] Kato, T., Kaneko, T., Sato, S., Nakamura, Y., & Tabata, S. (2000). Complete structure of the chloroplast genome of a legume, Lotus japonicus. *DNA Research*, 7, 323-330.

[97] Li, W. H. (1997). *Molecular Evolution*, Sunderland, Sinauer Associates, 487.

[98] Saski, C.h., Lee, S. B., Daniell, H., Wood, T. C., Tomkins, J., Kim, H. G., & Jansen, R. K. (2005). Complete chloroplast genome sequence of Glycine max and comparative analyses with other legume genomes. *Plant Molecular Biology*, 59, 309-322.

[99] Shinozaki, K., Ohme, M., Tanaka, M., Wakasugi, T., Hayashida, N., Matsubayashi, T., Zaita, N., Chunwongse, J., Obokata, J., Yamaguchi-Shinozaki, K., Ohto, C., Torazawa, K., Yamada, K., Kusuda, J., Takaiwa, F., Kato, A., Tohdoh, N., Shimada, H., & Sugiura, M. (1986). The complete nucleotide sequence of tobacco chloroplast genome: its gene organization and expression. *EMBO Journal*, 5, 2043-2049.

[100] Sugiura, M., & Shimada, H. (1991). Fine- structural features of the chloroplast genome- comparison of the sequenced chloroplast genomes. *Nucleic Acids Research*, 19, 983-995.

[101] Sugiura, M., & Sugita, M. (1992). The chloroplast genome. *Plant Molecular Biology*, 19, 149-168.

[102] Woźny, A. (1990). *Wykłady i ćwiczenia z biologii komórki*, Warszawa, Wydawnictwo Naukowe PWN.

[103] Palmer, J. D. (1985). Evolution of chloroplast and mitochondrial DNA in plants and algae. *In: MacIntyre RJ. (ed.) Monographs in Evolutionary Biology*, New York:, Molecular Evolutionary Genetics Plenum Press, 131-240.

[104] Wolfe, K.H. (1988). The site of deletion of the inverted repeat is pea chloroplast DNA contains duplicated gene fragments. *Current Genetic*, 13, 97-99.

[105] Palmer, J. D., Osorio, B., Aldrich, J., & Thompson, W. F. (1987). Chloroplast DNA evolution among legumes: loss of a large inverted repeat occurred prior to other sequence rearrangements. *Current Genetic*, 11, 275-286.

[106] Lavin, M., Doyle, J. J., & Palmer, J. D. (1990). Evolutionary significance of the loss of the chloroplast-DNA inverted repeat in the Leguminosae subfamily Papilionoideae. *Evolution*, 44, 390-402.

[107] Herdenberger, F., Pillay, D. T. N., & Steinmetz, A. (1990). Sequence of the trnH gene and the inverted repeat structure deletion of the broad bean chloroplast genome. *Nucleic Acids Research*, 18, 1297.

[108] Shaw, J., Lickey, E. B., Beck, J. T., Farmer, S. B., Liu, W., Miller, J., Siripun, K. C., Winder, C. T., Schilling, E. E., & Small, R. L. (2005). The tortoise and the here II: rela-

tive utility of 21 noncoding chloroplast DNA sequences for phylogenetic analysis. *American Journal of Botany*, 92, 142-166.

[109] Ravi, V., Khurana, J. P., Tyagi, A. K., & Khurana, P. (2008). An update on chloroplast genomes. *Plant Systematics and Evolution*, 271101-122.

[110] Kelchner, S. A. (2000). The evolution of non-coding chloroplast DNA and its application in plant systematics. *Annals of the Missouri Botanical Garden*, 87, 482-498.

[111] Perry, A. S., Brennan, S., Murphy, D. J., & Wolfe, K. H. (2006). Evolutionary re-organisation of a large operon in Adzuki bean chloroplast DNA caused by inverted repeat movement. *DNA Research*, 9, 157-162.

[112] Close, P. S., Shoemaker, R. C., & Keim, P. (1989). Distribution of restriction site polymorphism within the chloroplast genome of the genus Glycine, subgenus Soja. *Theoretical and Applied Genetics*, 77, 768-776.

[113] Xu, D. H., Abe, J., Kanazawa, A., Sakai, M., & Shimamoto, Y. (2000). Sequence variation of non-coding regions of chloroplast DNA of soybean and related wild species and its implications for the evolution of different chloroplast haplotypes. *TAG Theoretical and Applied Genetics*, 101, 5-6, 724-732.

[114] Kajita, T., Ohashi, H., Tateishi, Y., Bailey, C. D., & Doyle, J. J. (2001). rbcL and legume phylogeny, with particular reference to Phaseoleae, Millettieae, and allies. *Systematic Botany*, 26, 515-536.

[115] Nickret, D. L., & Patrick, J. A. (1998). The nuclear ribosomal DNA intergenic spacer of wild and cultivated soybean have low variation and cryptic subrepeats. *NRC Canada. Genome*, 41183-192.

[116] Palmer, J. D., Adams, K. L., Cho, Y., Parkinson, C. L., Qiu, Y. L., & Song, K. (2000). Dynamic evolution of plant mitochondrial genomes: mobile genes and introns and highly variable mutation rates. U.S.A., *Proceedings of the National Academy of Sciences*, 97, 6960-6966.

[117] Pennington, R. T., Klitgaard, B. B., Ireland, H., & Lavin, M. (2000). New insights into floral evolution of basal Papilionoideae from molecular phylogenies. *In: PS Herendeen A Bruneau (ed.) Advances in Legume Systematics Kew*, UK, 9, 233-248.

[118] Wojciechowski, M. F., Lavin, M., & Sanderson, M. J. (2004). A phylogeny of legumes (Leguminosae) based on analysis of the plastid matK gene resolves many well-supported subclades within the family. *American Journal of Botany*, 91, 1846-1862.

[119] Taberlet, P., Gielly, L., Pautou, G., & Bouvet, J. (1991). Universal primers for amplification of three non-coding regions of chloroplast DNA. *Plant Molecular Biology*, 17, 1105-1109.

[120] Wolfe, A. D., Elisens, W. J., Watson, L. E., & Depamphilis, C. W. (1997). Using restriction-site variation of PCR-amplified cpDNA genes for phylogenetic analysis of Tribe Cheloneae (Scrophulariaceae). *American Journal of Botany*, 84, 555-564.

[121] Sehgal, D., & Raina, S. N. (2008). DNA markers and germplasm resource diagnostics: new perspectives in crop improvement and conservation strategies. *In: Arya ID, Arya S. (ed.) Utilization of biotechnology in plant sciences*, Microsoft Printech (I) Pvt. Ltd., Dehradun, 39-54.

[122] Varshney, R. K., Graner, A., & Sorrells, M. E. (2005). Genic microsatellite markers in plants: features and applications. *Trends Biotechnology*, 23, 48-55.

[123] Varshney, R. K., Thudi, M., Aggarwal, R., & Börner, A. (2007). Genic molecular markers in plants: development and applications. *In: Varshney RK, Tuberosa R (ed.) Genomics assisted crop improvement: genomics approaches and platforms*, Springer, Dordrecht, 1, 13-29.

[124] Powell, W., Machray, G. C., & Provan, J. (1996). Polymorphism revealed by simple sequence repeats. *Trends Plant Sciences*, 1, 215-222.

[125] Russell, J. R., Fuller, J. D., Macaulay, M., Hatz, B. G., Jahoor, A., Powell, W., & Waugh, R. (1997). Direct comparison of levels of genetic variation among barley accessions detected by RFLPs, AFLPs, SSRs and RAPDs. *Theoretical and Applied Genetics*, 95, 714-722.

[126] Bohn, M., Utz, H. F., & Melchinger, A. E. (1999). Genetic similarities among winter wheat cultivars determined on the basis of RFLPs, AFLPs, and SSRs and their use for predicting progeny variance. *Crop Sciences*, 39, 228-237.

[127] Parida, S. K., Kalia, S. K., Sunita, K., Dalal, V., Hemaprabha, G., Selvi, A., Pandit, A., Singh, A., Gaikwad, K., Sharma, T. R., Srivastava, P. S., Singh, N. K., & Mohapatra, T. (2009). Informative genomic microsatellite markers for efficient genotyping applications in sugarcane. *Theoretical and Applied Genetics*, 118, 327-338.

[128] Shimamoto, Y., Hasegawa, A., Abe, J., Ohara, M., & Mikami, T. (1992). Glycine soja germplasm in Japan: isozymes and chloroplast DNA variation. *Soybean Genet Newsletter*, 19, 73-77.

[129] Xu, D. H., Abe, J., Gai, J. Y., & Shimamoto, Y. (2002). Diversity of chloroplast DNA SSRs in wild and cultivated soybeans: evidence for multiple origins of cultivated soybean. *Theoretical and Applied Genetics*, 105, 645-653.

[130] Xu, D. H., Abe, J., Kanazawa, A., Gai, Y., & Shimamoto, Y. (2001). Identification of sequence variations by PCR-RFLP and its application to the evaluation of cpDNA diversity in wild and cultivated soybeans. *Theoretical and Applied Genetics*, 102, 683-688.

[131] Hirata, T., Abe, J., & Shimamoto, Y. (1996). RFLPs of chloroplast and motochondrial genomes in summer and autumn maturing cultivar groups of soybean in Kyushu district of Japan. *Soybean Genetic Newsletter*, 23, 107-111.

[132] Shimamoto, A., Fukushi, H., Abe, J., Kanazawa, A., Gai, J., Gao, Z., & Xu, D. (1998). RFLPs of chloroplast and mitochondrial DNA in wild soybean, Glycine soja, growing in China. *Genetic Resources and Crop Evolution*, 45, 433-439.

[133] Chen, J., & Hebert, P. D. N. (1999). Directed termination of the polymerase chain re-action: kinetics and application in mutation detection. *Genome*, 42, 72-79.

[134] Orita, M., Suzuki, T. S., & Hayashi, K. (1989). Rapid and sensitive detection of point mutation and DNA polymorphisms using the polymerase chain reaction. *Genomics*, 5, 874-879.

[135] Sarkar, G., Yoon, H., & Sommer, S. S. (1995). Dideoxy fingerprinting (ddF): a rapid and efficient screen for the presence of mutations. *Genomics*, 13, 441-443.

[136] Manen, J. F., & Natall, A. (1995). Comparison of the evolution of ribulose- 1, 5-bi-sphosphate carboxylase (rbcL) and atpB-rbcL non-coding spacer sequences in a re-cent plant group, the tribe Rubieae (Rubiaceae). *Journal of Molecular Evolution*, 41, 920-927.

[137] Jordan, W. C., Courtney, M. W., & Neigel, J. E. (1996). Low levels of intraspecific ge-netic variation at a rapidly evolving chloroplast DNA locus in North American duck-weeds (Lemnaceae). *American Journal of Botany*, 83, 430-439.

[138] Sang, T., Crawford, D. J., & Stuessy, T. F. (1997). Chloroplast DNA phylogeny, reticu-late evolution, and biogeography of Paeonia (Paeoniaceae). *American Journal of Bot-any*, 84, 1120-1136.

[139] Small, R. L., Ryburn, J. A., Cronn, R. C., Seelanan, T., & Wendel, J. F. (1998). The tor-toise and the hare: choosing between noncoding plastome and nuclear Adh sequen-ces for phylogeny reconstruction in a recently diverged plant group. *American Journal of Botany*, 85, 1301-1315.

[140] Mc Dade, L. A., & Moody, M. L. (1999). Phylogenetic relationship among Acantha-ceae: evidence from noncoding trnL-trnFF chloroplast DNA sequences. *American Journal of Botany*, 86, 70-80.

[141] Molvray, M., Kores, P. J., & Chase, M. W. (1999). Phylogenetic relationships within Korthalsella (Viscaceaek) based on nuclear ITS and plastid trnL-F sequence data. *American Journal of Botany*, 86, 249-260.

[142] Demesure, B., Comps, B., & Petit, R. J. (1996). Chloroplast DNA phylogeography of the common beach (Fagus sylvatica L.) in Europe. *Evolution*, 50, 2515-2520.

[143] Wolfe, A. D., Elisens, W. J., Watson, L. E., & Depamphilis, C. W. (1997). Using restric-tion-site variation of PCR-amplified cpDNA genes for phylogenetic analysis of Tribe Cheloneae (Scrophulariaceae). *American Journal of Botany*, 84, 555-564.

[144] Asmussen, C. B., & Liston, A. (1998). Chloroplast DNA characters, phylogeny, and classification of Lathyrus (Fabaceae). *American Journal of Botany*, 85, 387-40.

[145] Cipriani, G., Testolin, R., & Gardner, R. (1998). Restriction-site variation of PCR-am-plified chloroplast DNA regions and its implication for the evolution of Actinidia. *Theoretical and Applied Genetics*, 96, 389-396.

[146] Friesen, N., Pollner, S., Bachmann, K., & Blattner, F. R. (1999). RAPDs and noncoding chloroplast DNA reveal a single origin of the cultivated Allium fistulosum from A. altaicum (Alliaceae). *American Journal of Botany*, 86, 554-562.

[147] Close, P. S., Shoemaker, R. C., & Keim, P. (1989). Distribution of restriction site polymorphism within the chloroplast genome of the genus Glycine, subgenus Soja. *Theoretical and Applied Genetics*, 77, 768-776.

[148] Kanazawa, A., Tozuka, A., & Shimamoto, Y. (1998). Sequence variation of chloroplast DNA that involves EcoRI and ClaI restriction site polymorphisms in soybean. *Genes Genetic Systems*, 73, 111-119.

[149] Downie, S. R., Olmstead, R. G., Zurawski, G., Soltis, D. E., Soltis, P. S., Watson, C., & Palmer, J. D. (1991). Six independents losses of the chloroplast DNA rpl2 intron in dicotyledons: Molecular and phylogenetic implications. *Evolutions*, 45, 1245-1259.

[150] Spielmann, A., & Stutz, E. (1983). Nucleotide sequence of soybean chloroplast DNA regions which contain the psbA and trnH genes and cover the ends of the large single-copy region and one end of the inverted repeats. *Nucleic Acids Research*, 11, 7157-7167.

[151] Sugiura, M., & Shimada, H. (1991). Fine- structural features of the chloroplast genome- comparison of the sequenced chloroplast genomes. *Nucleic Acids Research*, 19, 983-995.

[152] Taberlet, P., Coissac, E., Popanon, F., Gielly, L., Miquel, C., Valentini, A., Vermat, T., Corthier, G., Brochmann, C., & Willerslev, E. (2006). Power and limitations of the chloroplast trnL (UAA) intron for plant DNA barcoding. *Nucleic Acids Research*, 35(3).

[153] Palmer, J. D. (1991). Plastid chromosomes: Structure and Evolution. *In: Bogorag L., Vasil IK (ed.) Cell culture and somatic cell genetic of plants*, New York, Academic Press, 7, 5-53.

Permissions

All chapters in this book were first published by InTech Open; hereby published with permission under the Creative Commons Attribution License or equivalent. Every chapter published in this book has been scrutinized by our experts. Their significance has been extensively debated. The topics covered herein carry significant findings which will fuel the growth of the discipline. They may even be implemented as practical applications or may be referred to as a beginning point for another development.

The contributors of this book come from diverse backgrounds, making this book a truly international effort. This book will bring forth new frontiers with its revolutionizing research information and detailed analysis of the nascent developments around the world.

We would like to thank all the contributing authors for lending their expertise to make the book truly unique. They have played a crucial role in the development of this book. Without their invaluable contributions this book wouldn't have been possible. They have made vital efforts to compile up to date information on the varied aspects of this subject to make this book a valuable addition to the collection of many professionals and students.

This book was conceptualized with the vision of imparting up-to-date information and advanced data in this field. To ensure the same, a matchless editorial board was set up. Every individual on the board went through rigorous rounds of assessment to prove their worth. After which they invested a large part of their time researching and compiling the most relevant data for our readers.

The editorial board has been involved in producing this book since its inception. They have spent rigorous hours researching and exploring the diverse topics which have resulted in the successful publishing of this book. They have passed on their knowledge of decades through this book. To expedite this challenging task, the publisher supported the team at every step. A small team of assistant editors was also appointed to further simplify the editing procedure and attain best results for the readers.

Apart from the editorial board, the designing team has also invested a significant amount of their time in understanding the subject and creating the most relevant covers. They scrutinized every image to scout for the most suitable representation of the subject and create an appropriate cover for the book.

The publishing team has been an ardent support to the editorial, designing and production team. Their endless efforts to recruit the best for this project, has resulted in the accomplishment of this book. They are a veteran in the field of academics and their pool of knowledge is as vast as their experience in printing. Their expertise and guidance has proved useful at every step. Their uncompromising quality standards have made this book an exceptional effort. Their encouragement from time to time has been an inspiration for everyone.

The publisher and the editorial board hope that this book will prove to be a valuable piece of knowledge for researchers, students, practitioners and scholars across the globe.

List of Contributors

Adriana Gutiérrez-Díez, Enrique Ignacio Sánchez-González, Ivón M. Cerda-Hurtado and Ma. Del Carmen Ojeda-Zacarías
Universidad Autonoma de Nuevo Leon. Faculty of Agronomy General Escobedo, Nuevo Leon, Mexico

Jorge Ariel Torres-Castillo
Universidad Autonoma de Tamaulipas. Institute for Applied Ecology Ciudad Victoria, Tamaulipas, Mexico

A. Exadactylos
Department of Ichthyology and Aquatic Environment, School of Agricultural Sciences, Univ. of Thessaly, Volos, Hellas, Greece

Mei-Chen Tseng and Dai-Shion Hsiung
Department of Aquaculture, National Pingtung University of Science and Technology, Taiwan

Sanjuana Hernández-Delgado, Jairo Martínez-Mondragón, Homar R. Gill-Langarica and Netzahualcoyotl Mayek-Pérez
Centro de Biotecnología Genómica-Instituto Politécnico Nacional (CBG-IPN) Reynosa, Tamaulipas, Mexico

José S. Muruaga-Martínez
Instituto Nacional de Investigaciones Forestales, Agrícolas y Pecuarias (INIFAP) Texcoco, Mexico

Ma. Luisa Patricia Vargas-Vázquez
INIFAP-Texcoco, Mexico

José L. Chávez-Servia
CIIDIR-Unidad Oaxaca, IPN, Oaxaca Mexico

Andreea Dudu, Sergiu Emil Georgescu and Marieta Costache
University of Bucharest, Department of Biochemistry and Molecular Biology, Splaiul Independentei 91-95, Bucharest, Romania

G. Manzo-Sánchez, M.T. Buenrostro-Nava and S. Guzmán-González
School of Agronomy and Biological Sciences, University of Colima; Tecoman, Colima, Mexico

M. Orozco-Santos
INIFAP-Tecoman Experimental Station, Colima, Tecoman, Colima, Mexico

Muhammad Youssef
Department of Genetics, Faculty of Agriculture, Assiut University, Assiut, Egypt

Rosa Maria Escobedo-Gracia Medrano
Plant Biochemistry and Molecular Biology Unit, Yucatan Center for Scientific Research, Chuburna de Hidalgo, Merida, Yucatan, Mexico

João Carlos da Silva Dias
University of Lisbon, Instituto Superior de Agronomia, Tapada da Ajuda, Lisbon, Portugal

Ana María López and Marta Leonor Marulanda
Universidad Tecnológica de Pereira, Plant Biotechnology Laboratory, Colombia

Carlos Mario Ospina
Cenicafé, National Federation of Coffee Growers of Colombia, Colombia

Lidia Skuza, Ewa Filip and Izabela Szućko
Cell Biology Department, Faculty of Biology, University of Szczecin, Poland

Index

Printed in the USA
CPSIA information can be obtained
at www.ICGtesting.com
JSHW052310231023
50683JS00006BA/55